DNA Polymorphisms as Disease Markers

Edited by

D. J. Galton

St. Bartholomew's Hospital
London, United Kingdom

and

G. Assmann

Westfälische Wilhelms-Universität
Münster, Germany

Plenum Press
New York and London
Published in cooperation with NATO Scientific Affairs Division

Proceedings of a NATO Advanced Research Workshop
on DNA Polymorphisms as Disease Markers,
held September 26–28, 1990,
in London, United Kingdom

RB
147
. N37
1991

Library of Congress Cataloging in Publication Data

NATO Advanced Research Workshop on DNA Polymorphisms as Disease
Markers (1990: London, England)
 DNA polymorphisms as disease markers / edited by D. J. Galton and G.
Assmann.
   p.   cm.—(NATO ASI series. Series A, Life sciences; v. 214)
 "Proceedings of a NATO Advanced Research Workshop on DNA Polymor-
phisms as Disease Markers, held September 26–28, 1990, in London, United
Kingdom"—T.p. verso.
 Published in cooperation with NATO Scientific Affairs Division."
 Includes bibliographical references and index.
 ISBN 0-306-44039-3 (hardbound)
 1. Metabolism—Disorders—Genetic aspects—Congresses. 2. Metabolism—
Disorders—Etiology—Congresses. 3. Genetic polymorphisms—Congresses. 4.
Non-insulin-dependent diabetes—Genetic aspects—Congresses. 5. Hyper-
lipidemia—Genetic aspects—Congresses. 6. Atherosclerosis—Genetic Aspects—
Congresses.
I. Title. II. Series.
 [DNLM: 1. Diabetes Mellitus—genetics—congresses. 2. Atherosclerosis—
genetics—congresses. 3. DNA, Recombinant—congresses. 4. Genetic markers—
congresses. 5. Hyperlipidemia—genetics—congresses. 6. Polymorphism
(Genetics)—congresses. WK 810 N2778d   1990]
RB147.N37   1991
616.3'9042—dc20
DNLM/DLC   91-31838
for Library of Congress   CIP

ISBN 0-306-44039-3

© 1991 Plenum Press, New York
A Division of Plenum Publishing Corporation
233 Spring Street, New York, N.Y. 10013

Printed in the United States of America

DNA Polymorphisms as Disease Markers

NATO ASI Series

Advanced Science Institutes Series

A series presenting the results of activities sponsored by the NATO Science Committee, which aims at the dissemination of advanced scientific and technological knowledge, with a view to strengthening links between scientific communities.

The series is published by an international board of publishers in conjunction with the NATO Scientific Affairs Division

A	**Life Sciences**	Plenum Publishing Corporation
B	**Physics**	New York and London
C	**Mathematical and Physical Sciences**	Kluwer Academic Publishers
D	**Behavioral and Social Sciences**	Dordrecht, Boston, and London
E	**Applied Sciences**	
F	**Computer and Systems Sciences**	Springer-Verlag
G	**Ecological Sciences**	Berlin, Heidelberg, New York, London,
H	**Cell Biology**	Paris, Tokyo, Hong Kong, and Barcelona
I	**Global Environmental Change**	

Recent Volumes in this Series

Volume 209—Molecular Basis of Human Cancer
  edited by Claudio Nicolini

Volume 210—Woody Plant Biotechnology
  edited by M. R. Ahuja

Volume 211—Biophysics of Photoreceptors and Photomovements in Microorganisms
  edited by F. Lenci, F. Ghetti, G. Colombetti, D.-P. Häder, and Pill-Soon Song

Volume 212—Plant Molecular Biology 2
  edited by R. G. Herrmann and B. Larkins

Volume 213—The Midbrain Periaqueductal Gray Matter: Functional, Anatomical, and Neurochemical Organization
  edited by Antoine Depaulis and Richard Bandler

Volume 214—DNA Polymorphisms as Disease Markers
  edited by D. J. Galton and G. Assmann

Volume 215—Vaccines: Recent Trends and Progress
  edited by Gregory Gregoriadis, Anthony C. Allison, and George Poste

Series A: Life Sciences

PREFACE

The purpose of this workshop was to assess the value of DNA polymorphisms for the prediction, diagnosis or elucidation of aetiology for common metabolic diseases such as diabetes, hyperlipidaemia and atherosclerosis.

The advent of recombinant DNA techniques has produced an explosion in knowledge of restriction site polymorphisms and hypervariable sequences around candidate genes for such common metabolic diseases as atherosclerosis, hyperlipidaemia and diabetes mellitus. These diseases are the major causes of morbidity and mortality in Western societies today. Since 1983 it has become apparent that there is much more variation in the frequency and sites of these DNA polymorphisms in human populations than was initially appreciated and that the majority of these DNA polymorphisms are only linkage markers for the disease. Consequently it was considered timely for laboratories involved in the mapping of these DNA mutations to meet together to discuss the implications of their studies. The main issues are whether such DNA polymorphisms will lead to an identification of major aetiological loci and which are the best techniques to achieve this? What is the cause of the differences in frequencies of such polymorphisms amongst world populations? Are such studies best conducted in homogeneous populations or in pedigrees? Are haplotypes, though more laborious to construct, a better way to proceed than analysis of single site polymorphisms? What are the consensus polymorphic sites that relate to the common metabolic diseases of diabetes, atherosclerosis and the hyperlipidaemias?

The real need now to discuss the theoretical background behind the use of DNA polymorphisms and attempt to interpret all the information that has been obtained during the last seven years is incorporated in this volume. The importance of these studies is that they may lead directly to the identification of the inherited defects of the common metabolic diseases of diabetes, hyperlipidaemia and atherosclerosis and thus provide not only a better understanding of these diseases' pathologies and suggest new approaches to therapy but also identify genetic markers for early identification of individuals susceptible to these disorders.

D J Galton (London)
G Assmann (Munster)

CONTENTS

ATHEROSCLEROSIS

GENETIC APPROACHES TO COMMON DISEASES

Arno G. Motulsky

Departments of Medicine and Genetics
University of Washington
Seattle, WA 98195

ABSTRACT

Most common diseases have a genetic component. Multiple genes interacting with the environment are usually operative. Heterogeneity in etiology often occurs. Differences between monogenic and common diseases are outlined. Utilization of intermediate phenotypes rather than of the common disease *per se* as the diagnostic trait may be helpful for genetic analysis. Various genetic approaches with particular emphasis on linkage and association studies are discussed.

INTRODUCTION

The major advances of medical genetics have occurred in the unifactorial diseases. Examples include various unique chromosomal aberrations that produce an abnormal phenotype and the classic Mendelian disorders where the abnormal phenotype can be traced in pedigrees. Earlier, gene products were shown to be proteins or enzymes which could be mutationally altered in disease. Many enzyme defects were thereby explained. In more recent years came the ability to study DNA directly and to demonstrate the sites and specific nature of mutations (e.g., point mutation, deletion, frameshifts, etc.). Another development has been the recent availability of many DNA markers which can be used for studies of association linkage. This approach allows chromosomal mapping of monogenic diseases followed by elucidation of the specific defect. Cystic fibrosis is an example. Duchenne muscular dystrophy also has been clarified in this way and many other monogenic diseases are currently being studied in this manner.

COMMON DISEASES

In common familial disease (such as coronary heart disease) no single Mendelian genes are responsible in pathogenesis. Multiple genes are usually interacting and make analysis complex. The multiple loci contributing to the disease phenotype often cannot be studied directly since their mode of action is not clear. The only feasible approach to multiple gene action in the past therefore was statistical. A current working model assumes that among the multiple genes there is unequal contribution to the phenotype. One or a few major genes have a large effect and such major genes should be identifiable by current laboratory techniques. Other genes contribute less and may be considered as the "genetic background." Heterogeneity or the involvement of different

DNA Polymorphisms as Disease Markers, Edited by D.J. Galton and
G. Asmann, Plenum Press, New York, 1991

sets of genes in different pedigrees is a frequent complication that confounds analysis. The role of environmental factors in common disease is paramount. More work is needed to study environmental/genetic interactions as well as gene/gene interaction in the etiology of common disease.

Differences between monogenic and multifactorial common diseases

What are the differences between Mendelian monogenic diseases and the common multifactorial diseases? There are relatively few common diseases (diabetes, coronary heart disease, peptic ulcers, etc.) whereas there are several thousand of monogenic diseases that are individually rare. The age of onset of the monogenic diseases is usually early whereas the multifactorial diseases, because they require environmental interaction, are usually of late onset. The etiology of polygenic diseases is usually more heterogeneous than that of monogenic conditions. Risk prediction for multifactorial disease because of their complexity is less accurate but they are more amenable to prevention because of our ability to modify the involved environmental factors more readily.

Familial aggregation and twin studies

What methods for study of common diseases have been used? First came studies of familial aggregation, i.e., to find out if the disease occurred more frequently in family members as compared to controls. Familial aggregation can be due to sharing of a common environment that has nothing to do with genetics. To make sure that familial aggregation is not entirely environmental, correlation studies in different pairs of family members can be done. For example, among identical twins all genes are identical and the correlation coefficient for MZ twin pairs should be 1 if the trait under study is entirely genetic. Sib pairs, nonidentical twins, and parent-child pairs share on the average half of their parental genes. The correlation coefficient for such pairs should be 0.5 if the trait is entirely genetically determined. Correlation coefficients between adopted children and their parents who are not biologically related but live in the same environment can be used as controls. With identical twins, one should be aware that their environment is usually more alike than that of nonidentical twins since parents tend to treat monozygotic twins more similarly. Furthermore, identical twins because of their similarity tend to choose more similar environments. All these factors inflate MZ twin correlations. The most critical twin study approach to assess the role of heredity and environment is the study of identical twins reared apart, but such pairs are difficult to find. In the common diseases there is never 100% concordance of disease expression between identical twins so we know that nongenetic factors must be playing a part in pathogenesis.

Heritability

Heritability indices are often quoted in the literature. Heritability is the proportion of the total variance that is due to genetic factors, ranging from 100% or 1 (all genetic) to 0% (entirely environmental). This index is derived from twin and sib studies and is a population statistic. Heritability does not tell us anything about the mode of inheritance nor about the nature or number of genes involved. Heritability will vary with the environment so that if the environment is more alike, heritability values will be higher. The same set of genes in a population can therefore give different heritability indices with varying environments.

Approaches to analysis

How do we resolve possible heterogeneity? One approach is to find subtypes of the disease. Insulin-dependent diabetes and noninsulin-dependent diabetes were shown to have different genetic, physiological, and immunological causes.

The pathway from the various genes that are involved in a common disease to the disease phenotype is usually complex. It is therefore useful to study "intermediate phenotypes" rather than the disease phenotype itself for genetic analysis. "Intermediate phenotypes" can be identified by searching for genetic variation at the various pathophysiological steps that are involved in a disease. In coronary heart disease we know that hyperlipidemia is an important risk factor and therefore represents an intermediate phenotype. Since we know much about the biochemistry and physiology of the hyperlipidemias, we can analyze genetic variation affecting various aspects of the hyperlipidemias.

There are many heterozygotes (i.e., single dose gene carriers) for rare recessive diseases in the population. Since heterozygotes for various enzyme and receptor defects have only half of the functional activity compared to normal homozygotes, such carrier states may contribute to genetic susceptibility of common disorders. For example, in homocystinuria there are claims that heterozygotes have a higher incidence of peripheral and cerebrovascular disease.

Segregation and linkage analysis

In complex segregation analysis we ask which mode of inheritance is most likely to explain the family data at hand. A computer is programmed to examine the data from many families for the best fit to various modes of inheritance. As an example, segregation at a major locus against a polygenic background might be found to give the highest likelihood of explaining the data. The next step ideally is linkage analysis. The question asked here is whether there is cosegregation of a linkage marker with the disease or intermediate phenotype. This approach can be used often in view of the wealth of genetic markers arising from the new DNA technologies. RFLPs provided a great number of new genetic markers for use in linkage analysis. As many as one in 300 base pairs in flanking or intronic DNA may show variation and produce polymorphic sites. More recently a variable number of tandem repeats (VNTRs) have become useful. Other variation involving CA repeats have been found. There is so much variability at the DNA level that a great variety of markers spread over all the chromosomes are now available. Most families therefore are informative for linkage studies.

The most logical approach in linkage is to study candidate genes, i.e., genes that are known or at least are likely to be involved in the pathophysiology of the disease in question. DNA variation at the candidate gene loci can then be used as possible linkage markers. The candidate gene approach is less cumbersome than using random markers since one does not have to study several hundred of anonymous DNA markers spread over the entire genome. A linkage study therefore should start with a candidate gene. If no linkage can be found, the more laborious anonymous marker approach can be utilized.

How do we proceed practically? In most cases, a dominant disease that segregates in many individuals of a large family in multiple generations is most informative but it is often difficult to find such large kindreds. Smaller families can be combined for linkage study but raise problems of possible heterogeneity. Linkage analyses in diseases where no clear segregation is observed such as the common diseases pose analytical problems. For example, claims have been made for certain linkages in schizophrenia and manic depressive psychosis which could not be substantiated. Whether the problem is genetic heterogeneity, fuzzy phenotype assignments and the absence of intermediate phenotypes, or problems with the analysis is not yet clear. All these factors may have contributed to the current impasse.

ASSOCIATION STUDIES

Association studies were first done with the ABO blood groups and many different diseases. Much work was done with little impact on medicine. A few findings stood out. As an example, blood group O is more common in patients with peptic ulcer. The pathophysiology of this association is unknown.

Then came the HLA era. A strong association such as that of HLA B27 in ankylosing spondylitis was one result. The mechanism is unknown and probably not due

3

to linkage. In hemochromatosis, the mutant gene controls iron absorption and was found to be tightly linked to the HLA A locus after an HLA association was found. Among many autoimmune diseases (such as type I diabetes mellitus, multiple sclerosis, rheumatoid arthritis) associations with certain HLA types were detected. Linkage with immune determinants appears to be involved.

With DNA association studies in populations, we are looking for an increased frequency of a DNA marker in a disease or intermediate phenotypes as compared to controls. This usually assumes that there is a mutation that is closely linked to a DNA marker. Association of the marker with the disease or intermediate phenotype is searched for. To find a non-spurious association, the mutation should be identical in all those with the disease. If there is mutational heterogeneity, association with the DNA marker under study will be difficult to detect. The marker gene and the disease gene have to be tightly linked. They must display gametic association or linkage disequilibrium, i.e., they must be so tightly linked that over many generations they have remained together on the same chromosome and have not been separated by recombination. Racial and ethnic matching of disease and control groups is important. We know from various polymorphisms that there are differences in marker gene frequency between races and even between different populations of the same continent. If a study shows a statistically significant association between a marker and a disease, the finding may not necessarily have biological significance. A second series should be performed to see if the association still holds. Since a large number of comparisons with many markers are usually done in these studies, one must correct for multiple comparisons to rule out statistically significant but biologically meaningless associations. Associations are more likely to be found with intermediate phenotypes rather than with the clinical disease. For example, with coronary heart disease associations of DNA markers of lipid genes are more likely to be found with the hyperlipidemias rather than with coronary artery disease *per se* because factors other than hyperlipidemia determine the final outcome of coronary heart disease or myocardial infarction. Few of the many reported associations between DNA markers and disease have stood the test of time because of the above mentioned pitfalls. In the meantime it may be more useful particularly to employ more general risk predictors for coronary artery disease such as blood cholesterol (or possible apo B levels), Lp(a), and HDL levels. Their combined use already provides considerable information to assess risks.

Animal models

Animal models of disease may occasionally be useful. For example, there is much more homology between human genes and mouse genes than was previously suspected. For example, the diabetic genes in the NOD mouse has marked homology with the human genes at the HLA locus. Rodent models for hypertension and atherosclerosis exist. The significance of these models for the analgous human disorders remain questionable until it can be shown that the same gene or genes are involved in both humans and the roden species.

REFERENCES

For a more detailed coverage of various approaches to common disease see King RA, Rotter JF, and Motulsky AG: Chapter 1, The approach to genetic basis of common diseases in *The Genetic Basis of Common Disease*, Oxford U Press, New York/Oxford, 1991.

For molecular approaches see Childs B and Motulsky AG: Recombinant DNA analysis of multifactorial disease. Progr Med Genet 7:180-194, 1988.

A CANDIDATE GENE APPROACH TO THE GENETICS OF NORMAL LIPID VARIATION:

DNA ASSOCIATION STUDIES AND GENOTYPE BY ENVIRONMENT INTERACTION

Eric Boerwinkle[1] and Lawrence Chan[2]

[1]Center for Demographic and Population Genetics, The University
of Texas Health Science Center at Houston, Houston, TX 77225

[2]Departments of Cell Biology and Medicine, and the Diabetes and
Endocrinology Research Center, Baylor College of Medicine,
Houston, TX 77030

INTRODUCTION

The chronic diseases of later life such as diabetes, cancer and coronary heart disease represent the largest burden to public health. These complex diseases are the result of the interaction of numerous etiological factors, both genetic and environmental, each making a relatively small contribution to overall liability. Therefore, the contribution to disease risk of variation at any single gene is likely small when compared to all other risk factors combined. In addition, the contributing genes do not act alone; rather they are interacting with other genes and environmental factors in determining disease risk. Therefore, the effect of a gene measured in one environment or population may be very different from its effect in another.

Human molecular genetics have provided valuable advances in the prevention and understanding of several single gene disorders such as cystic fibrosis (Riordan et al, 1989). Several inborn errors of lipid metabolism have also been reported and characterized. For example, Brown and Goldstein (1989) and others have described lesions in the low density lipoprotein [LDL] receptor gene that lead to elevated LDL levels and premature atherosclerosis. Although these inborn errors of metabolism have a large impact on the affected individuals and families, their contribution to disease in the population is not large because of their relatively low frequency. Most of the genetic variability underlying lipid levels and disease risk is likely accounted for by a handful of polymorphic genes each having moderate effects.

In this paper, we briefly summarize select data indicating that genes contribute to the variability in lipid, lipoprotein, and apolipoprotein levels. We concentrate on the apolipoprotein genes in this summary. To date, the candidate apolipoprotein genes that have been analyzed belong to three groups: the soluble apolipoprotein gene family, which is comprised of apo A-I, A-IV, C-I, C-II, C-III, and E; apo(a) which belongs to the plasminogen gene family; and apo B which thus far appears to be unique. Using length variation in the signal peptide of the apo B gene as an example, we next discuss the contribution of the popular DNA association studies to our knowledge of the role of specific genes in determining lipid levels. We close with a review of some results pertaining to the interaction between genes and environmental factors as they combine to affect lipid, lipoprotein, and apolipoprotein levels.

ARE THERE GENES CONTRIBUTING TO LIPID LEVELS AND DISEASE RISK?

The role of genes in determining inter-individual variation in lipid levels and the familial aggregation of other risk factors is well established. Biometrical genetic studies using the correlation of lipid measures among family members have determined the relative contribution of genetic, household and random environmental factors to the variability in plasma lipid, lipoprotein, and

DNA Polymorphisms as Disease Markers, Edited by D.J. Galton and
G. Asmann, Plenum Press, New York, 1991

apolipoprotein levels. Table 1 summarizes the results of several such studies. For each phenotype, the contribution of genetic factors is substantial. For total cholesterol levels, over 50% of the variance is attributable to genes as compared with less than 10% for shared environmental effects. Biometrical genetic studies next asked whether there is evidence to support a role for a single gene with a major effect on risk factor levels. Evidence for a "major gene" is determined from the pattern of segregation of the trait in a sample of families or larger pedigrees (Morton and MacLean, 1974). A summary of the results of such segregation analyses on lipid, lipoprotein, and apolipoprotein levels is also contained in Table 1. There is significant evidence from a number of studies supporting the role of genes with large effects on lipid, lipoprotein, and apolipoprotein levels. After establishing that a major gene does exist, linkage analyses are often used to determine its location and possible identity. Linkage analyses have identified the genes responsible for major effects on two measures of lipid metabolism (Table 1).

Table 1

Contribution of genetic, common environmental, and individual specific effect to phenotypic variance

| | Effects | | | | |
Trait	Genes	Shared Environments	Individual Specific	Major Gene	Linkage
Total					
Cholesterol[a]	53%	7%	40%	Yes[d]	No
LDL-cholesterol[a]	52%	16%	32%	Yes[e]	Yes[j]
HDL-cholesterol[a]	46%	21%	33%	Yes[f]	No
Triglycerides[a]	52%	8%	40%	No	–
Apo B[b]	51%	14%	35%	Yes[g]	No
Apo A-I[b]	43%	5%	52%	Yes[h]	No
Lp(a)[c]	98%	–	2%	Yes[i]	Yes[k]

a. Iselius, 1988. b. Hamsten, 1986. c. Hasstedt and Williams, 1986. d. Moll et al, 1984. e. Morton et al, 1978. f. Hasstedt et al, 1986. g. Hasstedt et al, 1987. h. Moll et al, 1989. i. Morton et al, 1985. j. Leppert et al, 1986. k. Drayna et al, 1989.

WHAT ARE THE GENES CONTRIBUTING TO VARIATION IN LIPID LEVELS?

Although biometrical genetic studies indicate that genes are contributing to CHD risk factor levels, they do not lend information about the identity and role of specific genes. One does not know which genes are contributing to the total polygenic variance, nor the frequency of the contributing gene polymorphisms and their phenotypic effects. Knowledge concerning individual loci is necessary to identify individuals carrying specific mutations, and to characterize genetic factors involved in the disease process.

Recent advances in atherosclerosis research and molecular biology have identified many genes and gene products that are likely contributing to cardiovascular disease risk. A list of genes involved in lipid transport and metabolism is shown in Table 2. They have been studied by various laboratories and identified as the major candidate genes for lipid abnormalities and atherosclerosis. The molecular characteristics of each of these genes were reviewed recently (Chan and Dresel, 1990). The soluble plasma apolipoproteins A-I, A-II, A-IV, C-I, C-II, C-III, and E, all appear to belong to one multigene family (Li et al, 1988). They have similar genomic structures consisting of four exons and three introns located at highly analogous positions. The apo A-IV gene structure is slightly different in that it has lost its first intron. The similarity is most striking in the last 33 codons of exon 3 (exon 2 for apo A-IV) which has been termed the common block. This block contains three repeats of an eleven-codon unit and is present in all seven apolipoproteins. The last exon, in contrast, varies greatly in length, although it also contains repeat units of eleven or twenty-two codons. Interestingly, a lamprey apolipoprotein LAL1 (Pontes et al, 1987) and an avian apolipoprotein apo VLDL-II (Dugaiczyk et al, 1981, van het Schip et al, 1983) also have very similar primary and genomic structures, and probably belong to the same gene family. In contradiction to these apolipoproteins, apo D has completely different genomic and primary structures and is apparently not a member of this mutligene family.

Table 2

Genes for proteins involved in lipid transport and metabolism

I.	Apolipoprotein Genes	Chromosomal Location
	ApoA-I	11 (11q13-qter)
	ApoA-II	1 (1p21-qter)
	ApoA-IV	11 (11q13-qter)
	ApoB	1 (2p23-p24)
	ApoC-I	19 (19q)
	ApoC-II	19 (19q)
	ApoC-III	11 (11q13-qter)
	ApoD	3 (3q26.2-qter)
	ApoE	19 (19q)
	Apo(a)	6 (6q27)
II.	Lipoprotein-modifying Genes	
	CETP	16 (16q12-16q21)
	LCAT	16 (16q22)
	LPL	8 (8p22)
	HTGL	15 (15q21)
III.	Genes Related to Cellular Lipid Metabolism	
	LDL receptor	19 (19p)
	LDL receptor-related protein	12 (12q13-q14)
	HMGCoA reductase	5 (5q13.1-q14)
	HMGACoA synthetase	5 (5q14-p12)
	Liver FABP	2 (2p12-q11)
	Intestinal FABP	4 (4q28-q31)
	CRBP	3
	CRBP-II	3
	HSL	19 (19cent-q13.3)

Abbreviations are: CETP, cholesteryl ester transfer protein; LCAT, lecithin:cholesterol acyltransferase; LPL, lipoprotein lipase; HTGL, hepatic triglyceride lipase; HMGCoA, 3-hydroxyl-3-methylglutaryl coenzyme A; FABP, fatty acid binding protein; CRBP, cellular retinol binding protein; HSL, hormone-sensitive lipase.

Apo(a) and apo B-100 are the other major apolipoproteins distinct from those that are soluble. Apo(a) is the specific protein found in the Lp(a) lipoprotein particle. The structure of apo(a) was recently deduced from its cDNA sequence (McLean et al, 1987). The deduced amino acid sequence shows remarkable homology to plasminogen. The reported sequence of apo(a) contains 38 blocks of repeating units of approximately 114 amino acids. The first 37 repeats closely resemble kringle IV, and repeat 38 resembles kringle V of plasminogen. These repeats are followed by the protease domain which retains the His/Asp/Ser catalytic triad residues of the plasminogen protease domain.

Apo B-100 is the major protein in LDL where it constitutes the physiological ligand for the LDL receptor. Apo B-100 is also found in the lp(a) particle where it is covalently linked to apo(a) (Gaubatz et al, 1983). Although apo B-100 has some sequence similarity to the soluble apolipoproteins (DeLoof et al, 1987), its evolutionary relationship with the latter is unknown. Plasma concentrations of apo B-containing lipoproteins are important determinants of the propensity to develop atherosclerosis. Therefore, the apo B gene has been studied extensively as a major candidate gene in atherogenesis. A large number of apoB polymorphisms defined at the DNA, immunologic, and amino acid sequence levels have been identified (e.g. Boerwinkle et al, 1990). The functional significance of the vast majority of these polymorphisms is unknown. One sequence polymorphism consists of an Arg to Gln substitution at amino acid position 3500 (Soria et al, 1989). There is a strong association of this mutation with elevated plasma cholesterol, and individuals carrying this mutation are said to have the syndrome of familial defective apo B-100. Another polymorphism which potentially affects the biosynthetic pathway of apo B is one that changes the structure of its signal peptide (Boerwinkle and Chan, 1989; Boerwinkle et al, 1990). Eukaryotic signal peptides are important elements of most secretory proteins and are required for the translocation of the nascent polypeptide chain into the lumen of the rough endoplasmic reticulum (Randall and Hardy, 1989). The location of the apo B signal peptide polymorphism suggests that it may affect apo B synthesis and secretion.

The electrophoresed and ethidium bromide stained amplification products for each of the three apo B signal peptide alleles are shown in figure 1. The alleles and their amplification products were named according to the number of amino acid residues in the apo B signal peptide. In Caucasians, there were two alleles that differ by 3 amino acids [9 base pairs (bp)]. The amplification product of 5'ßSP-27, the larger Caucasian allele, was 93 bp and that of 5'ßSP-24, the smaller allele, was 84 bp. In Mexican-Americans a third allele was observed, 5'ßSP-29, whose amplification product was 99 bp in length. DNA sequence analyses indicated that the signal peptide alleles consist of the following: the longest allele (designated 5'ßSP-29) encodes 29 amino acids in the signal peptide and contains two copies of the sequence (CTG GCG CTG) encoding Leu-Ala-Leu, and a consecutive run of 8 (CTG) codons encoding 8 leucine residues; the medium sized allele (5'ßSP-27) encodes 27 amino acids and contains 2 copies of (CTG GCG CTG) but a run of only 6 (CTG) codons; the shortest allele (5'ßSP-24) encodes 24 amino acids and contains a single copy of (CTG GCG CTG) and a run of 6 (CTG) codons.

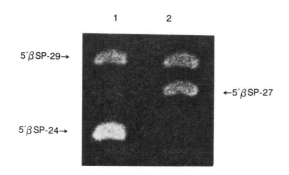

Figure 1. PCR products from each of the 3 apo B signal peptide alleles. Lane 1 contains a heterozygous 5'ßSP-24/29 individual. Lane 2 contains a heterozygous 5'ßSP-24/27 individual.

The estimated frequencies of the 5'ßSP-24 and 5'ßSP-27 apo B signal peptide alleles in a French Caucasian population were .355 and .645, respectively. The relative frequencies of the 5'ßSP-24, 5'ßSP-27, and 5'ßSP-29 alleles in the Mexican-American population were .337, .630, and .033, respectively. The frequencies of the 5'ßSP-24 and 27 alleles were not significantly different between Caucasians and Mexican-Americans.

We (Boerwinkle et al, 1991) have studied the association between the apo

B signal peptide polymorphism and altered lipid levels in two samples, Caucasians from Nancy, France and Mexican-Americans from Starr County, Texas. In a sample of French Caucasians, average apoA-I and glucose levels, and to a lesser extent, average triglyceride levels, were different among signal peptide genotypes (Table 3, panel A). Individuals homozygous for the 5'ßSP-24 allele had significantly higher plasma apoA-I levels than those with only one or no 5'ßSP-24 allele (1.59 mmol/l versus 1.42 mmol/l, respectively). Heterozygous individuals had average glucose levels that were significantly higher than the other two genotypes. Plasma triglyceride levels followed the same trend as apoA-I levels; homozygous 5'ßSP-24/24 individuals had higher levels than the other two genotypes. Average cholesterol, LDL-cholesterol and apo B levels were not different among signal peptide genotypes.

Table 3

A. Average lipid, apolipoprotein, glucose, and insulin levels
by apo B signal peptide genotype in a random sample from Nancy, France

Variable	5'ßSP-24/24	24/27	24/29	27/27	27/29	Total	$(\sqrt{MSE})^1$	p^2
N	15	110	0	72	0	197		
Cholesterol (mM)	5.72	5.96	–	5.90	–	5.92	(1.08)	(.93)
LDL-cholesterol (mM)	4.28	4.46	–	4.32	–	4.40	(1.13)	(.79)
Triglycerides (mM)	1.63	1.27	–	1.11	–	1.24	(0.93)	(.08)
HDL-cholesterol (mM)	1.46	1.27	–	1.36	–	1.32	(0.35)	(.14)
ApolipoproteinB (g/l)	1.23	1.20	–	1.14	–	1.18	(0.29)	(.29)
ApolipoproteinA-I(g/l)	1.59	1.41	–	1.45	–	1.44	(0.23)	(.03)
Glucose (mM)	5.23	5.65	–	5.38	–	5.52	(0.76)	(.05)

B. Average lipid, apolipoprotein, glucose, and insulin levels
by apo B signal peptide genotypes in a random sample from Starr County, Texas

Variable	5'ßSP-24/24	24/27	24/29	27/27	27/29	Total	$(\sqrt{MSE})^1$	p^2
N	17	84	4	68	8	181		
Cholesterol (mM)	4.88	5.10	5.35	5.02	5.17	5.06	(0.91)	.85 (.67)
LDL-cholesterol (mM)	3.21	3.15	4.14	3.13	2.73	4.14	(0.79)	.52 (.91)
Triglycerides (mM)	1.28	1.69	2.24	1.44	2.09	1.59	(0.99)	.19 (.17)
HDL-cholesterol (mM)	1.15	1.15	1.28	1.25	1.34	1.20	(0.27)	.11 (.08)
ApolipoproteinB (g/l)	.857	.937	1.187	.913	.949	.923	(0.21)	.46 (.53)
Apolipoprotein A-I (g/l)	1.25	1.12	1.31	1.17	1.23	1.16	(0.22)	.39 (.27)
Glucose (mM)	6.14	5.63	5.15	5.34	5.43	5.59	(1.02)	.04 (.01)
Insulin (pM)	102.6	104.4	134.4	90.0	112.2	99.9	(62.4)	.41 (.20)
C-peptide (nM)	1.00	1.09	1.50	0.91	0.72	1.01	(0.53)	.09 (.05)
Hemoglobin A1C (%)	6.6	5.9	6.0	6.1	5.6	6.1	(0.92)	.17 (.05)

1. Square root of the mean square error from the analyses of covariance. See the methods section for further details.
2. Probability of the equality of adjusted means considering all types and in parentheses considering only the 5'ßSP-24 and 27 alleles.

Results from a random sample of 188 Mexican-Americans from Starr County, Texas partially confirm the results obtained in the French Caucasian sample; plasma cholesterol, LDL-cholesterol and apo B levels were not significantly different among signal peptide genotypes (Table 3, panel B). In the random sample of Mexican-Americans average glucose levels were again significantly different among apo B signal peptide genotypes. However, the rank order of average glucose levels were not the same in the sample of Mexican-Americans and Caucasians. In the sample of Mexican-Americans, average plasma glucose levels were elevated in the 5'ßSP-24/24 homozygotes (6.14 mM) and lower in the other signal peptide genotypes. Blood levels of glycosylated hemoglobin, a monitor of integrated plasma glucose levels in the previous 3 months, were also elevated in the homozygous 5'ßSP-24/24 individuals. Plasma concentrations of C-peptide levels which have a longer half-life than insulin and are an index of pancreatic insulin reserve, were different among the common genotypes. Homozygous 5'ßSP-24/24 and heterozygous 5'ßSP-24/27 individuals had elevated C-peptide levels, and homozygous 5'ßSP-27/27 individuals had lower C-peptide levels. Insulin levels were not significantly different among genotypes. Mexican-Americans are a population with a high frequency of non-insulin dependent diabetes (Hanis et al, 1983). It is important to point out, though, that the 5'ßSP allele frequencies in this random sample of Mexican-Americans were not significantly different than a sample of diabetic Mexican-Americans (data not shown).

Xu et al (1990) reported that the apo B signal peptide genotypes were associated with plasma triglyceride levels in a sample of 106 individuals from North Karelia, Finland. In the sample of French Caucasians, the shorter 5'ßSP-24 allele was only weakly associated with triglyceride levels. Average triglyceride levels were not significantly different among genotypes in the random sample of Mexican-Americans. There are several possible reasons for these and other discrepancies encountered in DNA association studies. First, the significant results noted here may be due to chance. The second set of possible reasons for the observed discrepancy considers that the association is real, but different among populations. One cause of such a difference could be that there are other unidentified factors interacting with the apo B signal peptide and that these factors are different among populations. Another possibility, which we favor, considers that the signal peptide polymorphism was not directly causing the observed effect but rather it was in linkage disequilibrium with a second causal locus with a direct effect on glucose and triglyceride metabolism. In addition, the magnitude and direction of this disequilibrium was different among populations. We believe that design and analysis, along with consistency of results, be carefully considered when interpreting associations between DNA polymorphisms and lipid, lipoprotein, and apolipoprotein levels.

IS THERE EVIDENCE FOR GENOTYPE BY ENVIRONMENT INTERACTION?

Complex phenotypes such as total serum cholesterol levels are the result of the effects of numerous genetic and environmental factors. The gene for which effects of genetic variation on normal lipid levels are best understood is apo E. Apo E is a structural component of chylomicrons, very-low density lipoproteins [VLDL] and high-density lipoproteins [HDL] and is thought to have a major regulatory role in lipid metabolism via the cellular uptake of lipoproteins by both LDL and apo E specific receptors. Human apo E is polymorphic with three common alleles, $\epsilon2$, $\epsilon3$, and $\epsilon4$. The role of the apo E polymorphism in affecting lipid levels is well established. For example, in a sample of 563 unrelated Germans, the relative frequencies of the $\epsilon2$, $\epsilon3$, and $\epsilon4$ alleles were 0.063, 0.793, and 0.144, respectively. Average levels of total serum cholesterol, LDL-cholesterol, apo B, and apo E were significantly different among apo E genotypes. In this and most other populations, the average effect of the $\epsilon2$ allele is to significantly lower total cholesterol levels, and the average effect of the $\epsilon4$ allele is to significantly raise total cholesterol levels. These effects may be due to allelic differences in receptor binding, clearance of apo E bearing post-prandial chylomicrons and subsequent regulation of LDL receptors by intracellular cholesterol (Boerwinkle and Utermann, 1988).

Gueguen et al (1989) have investigated the interaction between the apo E polymorphism and changes in body weight as they affect the longitudinal profile of total cholesterol, triglyceride, ßlipoprotein, and glucose levels in a sample of 158 nuclear families from Nancy, France. There was no evidence for an effect of the apo E polymorphism on the longitudinal profile of any of the variables considered. This result was consistent with other studies that found constant effects of this gene across age groups (Boerwinkle and Sing, 1987; Davignon et al 1988; and Lehtimaki et al, 1990). However, there was a significant interaction between apo E effects and weight change on the longitudinal change of serum triglyceride and ßlipoprotein levels. Accompanying weight gain, individuals with an $\epsilon4$ allele showed a larger increase in triglyceride levels (0.15 +.03 mmol/l/Kg) compared to individuals with no $\epsilon4$ allele. The authors hypothesized that an increased production of VLDL as one gains weight along with retarded clearance of VLDL attributable to the effects of the $\epsilon4$ allele, may account for the observed interaction.

Tikkanen et al. (1990) and Xu et al. (1990) present results on the effects of the apo E polymorphism in a Finnish group that switched from their normal diet to a diet low in saturated fat. The frequency of the apo E alleles in this sample, and their effects while on a normal diet were similar to those previously reported in other populations. However, when switched to a low saturated fat diet, average total serum cholesterol levels were not significantly different among apo E genotypes. These results support the proposition of Boerwinkle and Utermann (1988), that an individual's response to changes in dietary cholesterol will differ among apo E genotypes. Boerwinkle and Utermann (1988) also suggest that the interaction between the apo E polymorphism and dietary fat intake will lead to differences in the magnitude of apo E effects among populations with different dietary habits. However, before concluding that dietary response differs among apo E genotypes, it is worth noting that a direct test of this hypothesis in the Finnish dietary intervention study did not detect significant differences (Xu et al, 1990). The relative decrease in total and LDL-cholesterol levels when individuals were placed on the low fat diet was not significantly different among apo E phenotypes (Figure 2).

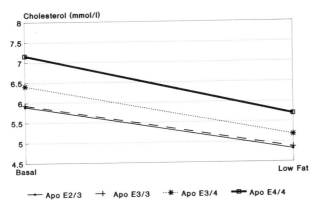

Figure 2. Average total serum cholesterol levels on each of two diets (a basal diet, and a diet low in total fat and saturated fat) for each of the common apo E genotypes in the Finnish dietary intervention study. The slopes of the lines connecting the two points indicate the change in serum cholesterol levels during dietary intervention (from Xu et al, 1990).

Although there is consensus that apo E does affect plasma cholesterol levels, given the extent of variation in plasma cholesterol and dietary cholesterol variation among populations, questions remain as to whether the biological effects of the apo E alleles on plasma cholesterol levels are the same in different populations. Despite the consistency in the direction of effects found in previous studies, it is difficult to asses whether the magnitude of effects is the same when comparing previously published results, since sampling, laboratory, and analytical methods often differ. We (Hallman et al, 1991) have compared the frequency and effects of the apo E polymorphism on total cholesterol levels in nine populations (Tyrolean, Sudanese, Indian, Chinese, Japanese, Hungarian, Icelandic, Finnish, and Malay). Apo E allele frequencies were significantly different among populations. The major apo E types in all populations were the E2/3 (frequency range: 7.0% in Indians to 16.9% in Malays), E3/3 (frequency range: 39.8% in Sudanese to 72.1% in Japanese), and E3/4 (frequency range: 11.3% in Japanese to 35.9% in Sudanese). The nine samples were grouped into six a priori categories and the technique of chi-square decomposition was used to investigate further the differences in apo E type frequency distribution (figure 3). Mean cholesterol levels were significantly different among the nine populations and ranged from a low of 144.2 mg/dl in the Sudanese to a high of 228.5 mg/dl in the Icelandics. Overall, the average effect of the $\epsilon 2$, $\epsilon 3$, and $\epsilon 4$ alleles on total cholesterol levels were -14.2, 0.04, and 8.14 mg/dl, respectively. Although there was considerable variation in this general trend, a two-way analysis of variance of the effects of population and apo E type on cholesterol levels showed no significant interaction effect, indicating that the influence of the apo E polymrophism on cholesterol levels did not differ significantly among populations.

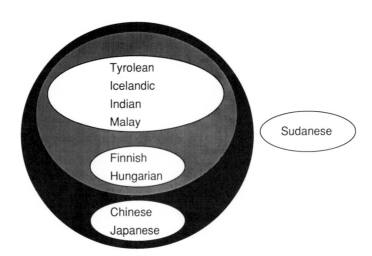

Figure 3. Graphical representation of the results of the chi-square decomposition for apo E allele frequencies among nine populations. Beginning with all of the populations, groups were split off and tested against the remaining populations (from Hallman et al, 1991).

CONCLUSIONS

In this review, we have posed three questions concerning genes, environment, and disease risk: "Are there genes contributing to lipid levels and disease risk?" "What are the genes contributing to variation in lipid levels?" and "Is there evidence for genotype by environment interaction?" The answer to the first question is generally accepted to be affirmative. Using a candidate gene approach, investigators are beginning to identify the gene loci, such as the apo E polymorphism, responsible for the familial aggregation of CHD risk factor levels. In answering the third question we have briefly reviewed the evidence for interaction between genes and environment for the apo B and apo E genes. Our understanding in this area is undergoing nearly continuous evolution. Although aided by the accessibility of considerable gene variation and the availability of appropriate analytical methods, progress is slowed by the innate contrariness of mankind and the imprecise nature of most environmental indices. Fueled by the shear magnitude of the problem and its public health implications, we expect continued, but slow, progress in this area in the coming years.

Acknowledgements

This work was supported by research grants HL-40613 to E.B., HL-34823 to C.L.H., and HL-27341, DK-27685 Baylor Diabetes and Endocrinology Research Center and The March of Dimes Birth Defects Foundation to L.C. The authors would like to thank Drs. Gerd Utermann, Craig L. Hanis, and Wen-Hsiung Li for their valuable collaborative assistance during the completion of various aspects of the work presented in this review.

References

Betteridge DF. Lipoprotein Metabolism. In: Recent Advances in Diabetes. M Nattrass M, Ed. Livingston, NY, Churchill, 1986, p. 91-107.

Boerwinkle E and Sing CF. The use of measured genotype information in the analysis of quantitative phenotypes in man. III. Simultaneous estimatation of the frequency and effects of the apolipoprotein E polymorphism and residual polygenetic effects on cholesterol, betalipoprotein, and triglyceride levels. Ann. Hum. Genet. 51: 211-226.

Boerwinkle E and Utermann G. Simultaneous effects of the apolipoprotein E polymorphism on apolipoprotein E, apolipoprotein B, and cholesterol metabolism. Am. J. Hum. Genet. 42:104-112, 1988.

Boerwinkle E and Chan L. A three codon insertion/deletion polymorphism in the signal peptide region of the human apolipoprotein B (Apo B) gene directly typed by the polymerase chain reaction. Nucl. Acids Res 17:4003, 1989.

Boerwinkle E, Hanis CL and Chan L. A unique length polymorphism in the signal peptide region of the apolipoprotein B gene in Mexican-Americans. Nucl. Acids Res. 18: 7193, 1990.

Boerwinkle E, Lee SS, Butler R, Schumaker VN, Chan L. Rapid typing of apolipoprotein B DNA polymorphisms by DNA amplification. Atherosclerosis 1990, 81:225-232.

Boerwinkle E, Chen W-H, Visvikis S, Hanis CL, Siest G, and Chan L. Signal peptide length variation in the human apolipoprotein B gene: Molecular characteristics and association with plasma glucose levels. Diabetes (submitted).

Brown MS and Goldstein JL. Familial hypercholesterolemia. In: The metabolic basis of inherited disease 1989, (Scriver CR, Beaudet AL, Sly WS, and Valle D eds) pp 1215-1250.

Chan L and Dresel HA. Genetic factors influencing lipoprotein structure: Implications for atherogenesis. Lab. Invest. 1990, 62: 522-537.

Davignon J, Gregg RE, and Sing CF. Apolipoprotein E polymorphism and atherosclerosis. Arteriosclerosis 1988, 8:1-21.

Drayna DT, Hegele RA, Hass PE, Emi M, Wu LL, Eaton DL, Lawn RM, Williams RR, White RL, and Lalouel J-M: Genetic linkage between lipoprotein(a) phenotype and a DNA polymorphism in the plasminogen gene. Genomic 1989, 3:230-236.

Dugaiczyk A, Inglis AS, Strike PM, Burley RW, Beattie WG, Chan L. Comparison of the nucleotide sequence of cloned DNA coding for an apoprotein (apoVLDL-II) form avian blood and the amino acid sequence of an egg yolk protein (apovitellenin I): equivalence of the two sequences. Gene 1981, 14: 175-182.

Gaubatz JW, Heideman C, Gotto AM, Morrisett JD, and Dahlen GH. Isolation and characterization of the two major apoproteins in human lipoprotein(a). J. Biol. Chem. 1983, 258:4582-4589.

Gueguen R, Visvikis S, Steinmetz J, Siest G, and Boerwinkle E: An analy sis of genotype effects and their interactions by using the apolipoprotein E polymorphism and longitudinal data. Am. J. Hum. Genet. 1989, 45:793-802.

Hallman DM, Boerwinkle E, Saha N, Sandholzer C, Menzel HJ, Csazar A, Utermann G. The apolipoprotein E polymorphism: A comparison of allele frequencies and effects in nine populations. Am. J. Hum. Genet. (Submitted).

Hamsten A, Iselius L Dahlen G, de Faire U. Genetic and cultural inheritance of serum lipids, low- and high-density lipoprotein cholesterol and serum apolipoprotiens A-I, A-II, and B. Atherosclerosis 60:199-208, 1986.

Hanis CL, Ferrell RE, Barton SA, Aguilar L, Garza-Ibarra A, Tulloch BR, Garcia CA, Schull WJ: Diabetes among Mexican Americans in Starr County, Texas. Am. J. Epidemiol. 118: 659-672, 1983.

Hasstedt SJ, Ash KO and Williams RR. A reexamination of the major locus hypothesis for high density lipoprotein cholesterol level using 2170 persons screened in 55 Utah pedigrees. Am. J. Med. Genet. 24:57-66, 1986.

Hasstedt SJ and Williams RR. Three alleles for quantitative lp(a). Genet. Epid. 3:53-55, 1986.

Hasstedt SJ, Wu L, and Williams RR. Major locus inheritance of apolipoprotein B in Utah pedigrees. Genet. Epid. 4:67-76, 1987.

Iselius L. Genetic epidemiology of common diseases in humans. In: Quantitative Genetics (Weir BS, Eisen EJ, Goodman MM and Namkoong G, eds.) pp 341-352, 1988.

Lehtimaki T, Moilanen T, Viikari J, Akerblom HK, Ehnholm C, Ronnemaa T, Marniemi J, Dahlen G, and Nikkari T. Apolipoprotein E phenotypes in Finnish youths: A cross-sectional and 6-year follow-up study. J. of Lipid Res. 31:487-485, 1990.

Leppert MF, Hasstedt SJ, Holm T, O'Connell P, Wu L, Ash O, Williams RR, and White RR. A DNA probe for the LDL receptor gene is tightly linked to hypercholesterolemia in a pedigree with early coronary heart disease. Am. J. Hum. Genet. 39:300-306, 1986.

Li W-H, Tanimura M, Luo C-C, Datta S, and Chan L. The apolipoprotein multigene family: biosynthesis, structure, structure-function relationships and evolution. J. Lipid Res. 1988, 29: 245-271.

Moll PP, Berry TD, Weidman WH, Ellefson R, Gordon H, and Kottke BA. Detection of genetic heterogeneity among pedigrees through complex segregation analysis: An application to hypercholesterolemia. Am. J. Hum. Genet. 36:197-211, 1984.

McLean JW, Tomlinson JE, Kuang W-J, Eaton DL, Chen EY, Fless GM, Scanu AM, Lawn RM. cDNA sequence of human apolipoprotein (a) is homologous to plasminogen. Nature 1987, 330: 132-137.

Moll PP, Michels VV, Weidman WH, and Kottke BA: Genetic determination of plasma apolipoprotein AI in a population-based sample. Am. J. Hum. Genet 44:124-139, 1989.

Morton NE, Gulbrandsen CL, Rhoads GG, Kagan A, and Lew R. Major locus for lipoprotien concentrations. Am. J. Hum. Genet. 30:583-589, 1978.

Morton NE and MacLean CJ. Analysis of family resemblance. III. Compelx segregation of quantitative traits. Am. J. Hum. Genet. 26:489-503, 1974.

Morton NE, Berg K, Dahlen G, Ferrell RE, and Rhoads G. Genetics of the lp lipoprotein in Japnaese-Americans. Genet. Epid. 2:113-121, 1985.

Pontes M, Xu X, Graham D, Riley M, Doolitle RF. cDNA sequences of to apolipoproteins form lamprey. Biochemistry 1987, 26:1611-1617.

Randall LL, Hardy SJS: Unity in function in the absence of consensus in sequence: Role of leader peptides in export. Science 243: 1156-1159, 1989.

Riordan JR, Rommens JM, Kerem B, Alon N, Rozmahel R, Grzelczak Z, Zielenski J, Lok S, Plavsic N, Chou J-L, Drumm M, Iannuzzi MC, Collins FS, Tsui L-C. Identification of the cystic fibrosis gene: Cloning and characterization of complementary DNA. Science 245:1066-1073, 1989.

Soria LF, Ludwig EH, Clarke HRG, Vega GL, Grundy SM, McCarthy BJ (1989) Association between a specific apolipoprotein B mutation and familial defective apolipoprotein B-100. Proc. Natl. Acad. Sci. 1989, 86:587-591.

Tikkanen MJ, Huttunen JK, Ehnholm C, and Pietinen P. Apolipoprotein E4 homozygosity predisposes to serum cholesterol elevation during high-fat diet. Arteriosclerosis 10:285-288, 1990.

van het Schip A, Meijlink RCRW, Stryker R, Gruber M, van Vliet AJ, van de Klundert AM, AB G. The nucleotide sequence of the chicken apo very low density lipoprotein gene. 1983 Nucl. Acids Res. 11:2529-2540.

Xu C-F, Boerwinkle E, Tikkanen MJ, Huttunen JK, Humphries SE, and Talmud PJ. Genetic variation at the apolipoprotein gene loci contribute to response of plasma lipids to dietary change. Genet. Epi. 7: 261-275, 1990.

Xu C-F, Tikkanen MJ, Huttunen JK, Pietinen P, Butler R, Humphries S, and Talmud P: Apolipoprotein B signal peptide insertion/deletion polymorphism is associated with Ag epitopes and involved in the determination of serum triglyceride levels. J. Lipid Res. 31: 1255-1261, 1990.

14

MOLECULAR GENETICS APPROACH TO POLYGENIC DISEASE— INITIAL RESULTS FROM ATHEROSCLEROSIS RESEARCH

Harald Funke, Arnold von Eckardstein, and Gerd Assmann

Institut für Klinische Chemie und Laboratoriumsmedizin
und Institut für Arterioskleroseforschung an der WWU
Westfälische Wilhelms-Universität Münster
Albert-Schweitzer-Str. 33
D-4400 Münster, F.R.G.

Introductory Remarks

Progress in the development of molecular genetics techniques has led in recent years to the identification of a variety of basic defects in genetic disease. This success in understanding inborn errors of metabolism, however, has been largely restricted to monogenic disease or to defects that involve large aberrations in the DNA primary structure. A much more complicated situation is present in common disease where important roles for disease expression have been attributed to endogenous as well as exogenous factors (1).

Among these diseases atherosclerosis is one of the leading causes for physical inability and death in Western World countries (2). It is characterized by the presence of a morphological substrate, called plaque, in the arterial wall that reduces oxygen supply to dependent structures. These plaques are formed with increasing age in most individuals. Therefore, it is important to separate out factors that accelerate or newly initiate this process and thereby cause a disease onset at an unusual early age. Our current knowledge on the formation of atherosclerosis is restricted to the identification of a number of factors that when present in an individual put him or her at a higher risk to develop early clinical manifestations of the disease. Among these risk factors which include high blood pressure, diabetes mellitus, lipoprotein metabolism anomalies, cigarette smoking,

DNA Polymorphisms as Disease Markers, Edited by D.J. Galton and
G. Asmann, Plenum Press, New York, 1991

family history of myocardial infarction, clinical symptoms of the disease, and age some have a suspected or proven genetic background and others are of exogenous origin (3). But even outside factors, like cigarette smoking, may have a strong genetic component through the presence or absence of "tolerance genes".

The presence of more than one risk factor in one individual has been shown to further increase the likelihood of disease development. This suggests that defects in many different genes and likely also in different metabolic pathways may contribute to pathological processes whose morphological consequence is the atherosclerotic plaque. It is currently not known if there is a common pathomechanism that is initiated by all the different defects underlying the various risk phenotypes or if plaque formation is the result of independent interactions of these factors with the arterial wall. Our knowledge about mechanisms that trigger or repair morphological changes in the arterial wall is only beginning to develop.

Epidemiological data of recent years have identified hypercholesterolemia and HDL deficiency to range among the most predictive single parameter risk indicators for atherosclerosis. Therefore it is of interest to better understand the factors that contribute to the formation of these biochemical conditions many of whom fulfill the characteristics of a quantitative intermediate trait for atherosclerosis.

What is the right strategy ?

The question which would be the right strategy for the disclosure of maior genetic components of atherosclerosis is difficult and challenging at the same time. Currently we have no idea what to expect with regard to the number of genes involved. It could be anything ranging from the effect of a few maior genes to a network of interrelations between a multitude of maior, intermediate, and minor genes. There are data suggesting that for the development of some quantitative intermediate phenotypes, namely excessive hypercholesterolemia and HDL deficiency, the latter might be the more likely hypothesis.

One of the most successful stories in atherosclerosis research has been the identification of defects in the LDL-receptor gene as causative agents for a substantial fraction of early onset atherosclerosis (4). Mutations in this gene currently provide the only well documented example for a maior gene effect on both the formation of a quantitative intermediate phenotype and progression towards disease development. Moreover, it is an excellent example for the successful application of the top-down approach to common disease.

In cases, however, that do not show such a strong association with disease and a clearly identifyable intermediate phenotype the bottom-up strategy may be what one is left with. Specifically, when the value of a quantitative intermediate trait is doubled by the heterozygous presence of a single gene defect, as is the case with most known LDL-receptor defects, a reliable phenotype assignment is still possible even in the presence of a modulating influence of mutations in other genes. This is not possible when the phenotypic effect of a mutation is small. Consequently, minor effects on a quantitative trait were found only after the identification of the underlying mutation (5,6). Interestingly, many defects in minor genes have a much higher frequency than those found in maior genes so that despite a small influence on the phenotype the overall effect on the population may be greater for a minor gene.

A very recent publication on the identification of linkage between the APOLP1 locus on chromosome 11 and familial combined hyperlipidemia (FCH) (7) is an example for the combination of the specific advantages of the top-down and the bottom-up approaches where problems in phenotype assignment were overcome by a very restrictive definition of FCH-carriers.

Another important step towards the identification of the genetic architecture of atherosclerosis was made through the analysis of the consequences that arise from two mutations with known phenotypic effects when they are present in the same individual (8,9).

Clearly, at present there is no one-works-for-all strategy to the genetic basis of common disease.

What is the right method ?

This question is as difficult as the one for the right strategy. Table I lists the methods we have used in our laboratory. In the following we give examples from research in our laboratory for some of the statements made in the table.

Table 1. Advantages and disadvantages of various methods used in the identification of genetic defects contributing to complex disease

	Association Analysis	Cosegregation Analysis	Candidate Gene Sequencing
Principle	statistical analysis of prevalence of allelic markers in defined groups	tests coordinate inheritance of clinically or biochemically defined phenotype and markers at the candidate gene locus	candidate genes are sequenced completely and their sequence is compared to wild type sequence; differences are evaluated by statistical and/or biochemical methods
Requirements	defined allelic markers at known or unknown loci	dominantly or codominantly inherited phenotype; allelic test loci; VNTRs ?	primary structure of candidate gene
Advantages	can recognize small contributions of a locus to phenotype variance; works best with markers that alter encoded protein sequence	negative results exclude candidate locus; easy to use diagnostic tool in diseases with high penetrance and expressivity	works with small phenotypic variance; allows to use the (suspected) defect itself as a marker for further studies
Drawbacks	small effects need large sample sizes; effects may not be seen at all because of linkage disequilibrium; allelic diseases are often not recognized	false negative results when there are difficulties in phenotype assessment; problems with quantitative phenotypes; not advisable in poly genic disease	work intensive; every mutation that can cause a biochemical consequence must be followed; use with large genes needs methods improvements
Verification	statistical genetics and / or expression systems	statistical genetics and / or expression systems	statistical genetics and / or expression systems
Positive examples	apo E4 (hypercholesterolemia) apo E2 (hypocholesterolemia)	LDL-receptor; classic genetic disease	LCAT; LPL; CETP; apo A-I; apo A-IV; apo C-III; apo B-3500

18

Allele	Location	Frequency	
		MI-Patients	Students
X2 (Xmnl)	.5'A-I	0.18	0.12
D2 (Del)	.5'A-I	0.05	0.05
A2 (Apal)	.5'A-I	0.40	0.35
M2 (Mspl)	.exon 3 of A-I	0.10	0.10
M3 (Mspl)	.exon 1 of A-I	0.04	0.05
P2 (Pstl)	.3'A-I	0.03	0.06
S2 (Sstl)	.3'C-III	0.12	0.10
V2 (Pvull)	.exon 1 of C-III	0.24	0.24

Data were determined in 314 patients and 207 students

Figure 1. Frequency distribution of alleles defined by dimorphic RFLP markers at the APOLP1 locus.

Allel number	Haplotypes X D A M M P S V	Frequency (segregation analysis only)
1	- - - - - - - -	0.50
2	- - - - - - - +	0.20
3	+ - + - - - - -	0.10
4	- - + - - - - -	0.03
5	- - + + - - + -	0.08
6	- + + - + - - +	0.02
7	- - + - - + - -	0.03
8	- - - - - + - -	<0.02
9	- - - - - - + -	<0.02
10	- - + - - + - +	<0.02
11	- + + - - - - -	0.02
12	- - + - - - - +	<0.02
13	+ - - - - - - -	<0.02
14	- - - - + - - +	<0.02
15	- - + + - + + -	-
16	- - + + - + + +	-
17	- + + + - - + -	-
18	+ - + - - + - -	-

Figure 2. Haplotype frequency deduced from the analysis of 68 individuals from 17 families.

Association analysis

This type of analysis is based on the determination of the prevalence of allelic markers in different groups and is thus well suited for case control studies. For ethical as well as for practical reasons the determination of an atherosclerosis phenotype is restricted to diseased patients. This method allows to test directly a marker´s association with atherosclerosis while other methods can only use the easier to assess intermediate phenotypes.

Restriction fragment length polymorphisms (RFLPs) are easy to determine markers that are found at nearly any candidate locus. This explains their frequent use in case control studies. Among the most widely used markers in atherosclerosis research are RFLPs at the APOLP1 locus. There are reports on an association of a Sst I RFLP with hypertriglyceridemia (10) or with myocardial infarction (11). Others have reported that a Pst I RFLP was associated with HDL deficiency (12). Our own results (fig.1) show a significant difference in the allele frequency between a student control group and MI-patients with angiographically verified coronary artery disease (CAD) only for the X2 allele of a Xmn I polymorhism that is located 3´ of the apo A-I gene. Genotype analysis was done on the same subjects in an attempt to single out a X2-containing haplotype that was more strongly associated with CAD. A haplotype frequency table for the population (fig.2) was used to transform genotypes into haplotypes. None of the such identified haplotypes showed significantly different frequencies in the two groups. In an explorative test we have analysed the presence of associations between any of the eight RFLPs and 7 different lipid and lipoprotein plasma concentrations. Several associations between RFLPs and lipid or lipoprotein parameters we found to be statistically significant. However, when a different statistical method (Kruskal/Wallis instead of Whitney/Mann) was applied to data analysis, previously significant associations disappeared, while others were newly identified. This problem occurs because the first test the phenotype associated with homozygosity for the rare allele is overestimated while in the other it is completely neglcted. An additional problem was that the identified associations were not stable between sexes and between the case and control group.

Although we have not found convincing associations between markers at the APOLP1 locus and CAD or lipid and lipoprotein concentrations from these results it cannot be concluded that this locus is irrelevant to either of the phenotypes. On the contrary, at least ten different defects have been identified at this locus that cause a reduction in plasma HDL concentrations (reviewed in 14). Three of these have been reported to be associated with also with CAD. Also, FCH has been linked to this locus (7). Together, these findings demonstrate that association analysis is not well suited for the identification of alleleic defects with a low frequency.

There are various reasons for the conflicting association results obtained by different groups. One obvious reason is the different ethnical composition of the groups studied by different investigators. In this context, though, the similarity of the haplotype frequencies we have observed in the Westfalian population (fig.2) and those that have been reported for an American population (15) is quite remarkable. Other causes for the heterogeneity of results are related to the statistical analysis. As documented by the above mentioned example explorative data analysis is not suited for documenting the existence of associations. Often the case numbers of the studies are simply too low.

False negative results can be obtained for highly allelic gene defects such as the LDL-receptor. A related reason includes phase disruptions between the marker and the functionally active defect. We have used haplotypes constructed from three different dimorphic RFLPs in the apolipoprotein C-II gene, which is located approximately 40 kb downstream of the apolipoprotein E gene, to identify the well established plasma cholesterol elevating effect of a point mutation in the apolipoprotein E gene (apo E4). None of the frequent C-II haplotypes showed an association with hypercholesterolemia. In fact, when the haplotypes were extended to include the apo E mutation three different haplotypes with this mutation have been identified. Thus, in association analysis it may be necessary to use a marker that is located very close to the defective site. Indeed, associations that have been found existent in different populations were often based on the identity of marker and defect.

Cosegregation analysis

In cosegregation analysis the coordinate inheritance of a given phenotype and markers at the candidate gene locus is tested. It is a special case of linkage analysis in which the assumed recombination fraction Θ is zero. It is suited for the analysis of dominantly or codominantly inherited phenotypes. One other requirement is an undisputed phenotype assignment, as any misclassification would have deleterious effects on the LOD score, especially when small families are analysed.

Two examples from our laboratory show that the influence of as yet unidentidied genes or unknown exogenous factors may lead to an incorrect phenotype assignment. The first example relates to the HDL-lowering effect of a point mutation in the apolipoprotein A-I gene that was observed in four unrelated families (16). Nine out of twelve heterozygote carriers of this mutation that causes an Arg for Pro exchange at position 165 of the mature apo A-I protein had HDL-cholesterol plasma concentrations below the 25th percentile. The phenotype assigned to one heterozygote whose HDL-cholesterol was near the 65th percentile would clearly have been normal. Using cosegregation analysis the hypothesis of an association of a defect at the APOLP1 locus with the HDL-deficient phenotype would thus have been dismissed. The second example is provided by a point mutation in the LCAT gene that, when present in homozygous form, causes fish-eye disease (FED) (17). Homozygous FED carriers´ plasma is devoid of HDL-cholesterol, while heterozygotes show a significant reduction compared to normal. However, only four out of nine FED heterozygotes showed HDL-cholesterol concentrations below the 25th percentile which again shows that phenotype assignments in quantitative traits are complicated. Moreover, in this example HDL deficiency is a phenotype that is secondary to impaired LCAT function. Impaired specific activity of LCAT is the, 100 % expressed, primary phenotypic consequence of the FED mutation. Thus, in cosegregation analysis there are problems of reduced penetrance and variable expression that may relate to the analysis of a phenotype that is only secondary or tertiary to the basic mutation.

These requirements render this method not specifically attractive for the analysis of quantitative traits. However, as mentioned before, a very rigid phenotype definition combined with the exclusion of genetic factors brought into the families by spouses led to the identification of linkage of FCH to APOLP1. It is not currently known if this restriction in phenotype definition may at the same time restrict the population relevance of the finding.

In conclusion, the cosegregation analysis approach is a tool that has been successfully used in the identification of defective loci underlying quantitative traits. One should , however, be aware of the many problems that may arise from its use.

Candidate gene sequencing

In all cases where a candidate gene has been established by independent methods, such as linkage analysis, biochemical data or purely guesswork,

sequence analysis of the gene under consideration may help to identify gene defects that are causative of a specific phenotype.

In our laboratory this approach has been successfully used in the identification of i) several defects in the apolipoprotein A-I gene that underly HDL-deficiencies (6, 18, and unpublished observations), ii) mutations in the apolipoprotein E gene that cause abnormal physicochemical behaviour of the encoded gene products (19 and unpublished observations), iii) one basic defect in fish-eye disease (17), iv) basic defects in lipoprotein lipase deficiency (20 and unpublished observation), and v) several cases of LCAT deficiency (21).

Gene sequencing can also be used as a tool in a consequent bottom-up strategy in which the consequences of mutations for phenotype expression are analysed subsequent their identification. In our laboratory we have used this strategy for the identification of mutations in the genes coding for apolipoproteins A-IV and E. Some of the such identified mutations showed small phenotypic effects on plasma lipid concentrations while others only showed alterations in the physicochemical behaviour of the encoded protein and still others were found to be without any apparent associated effect. Thus, this method is also a useful tool for the assignment of functinal domains to proteins.

One of the maior drawbacks of this method is the requirement for a known primary structure of the candidate gene. In large genes additional problems will arise from the high probability for the presence of more than one deviation from the wild type sequence.

Concluding remarks

We have demonstrated that with the combined use of the three above described methods the identification was possible of some genetic defects contributing to phenotypic expression in multifactorial disease. There is, however, the need to further improve these methods and to develop new techniques to be adequately prepared for the enormous challenge of unravelling most of the genetic components important in atherosclerosis development.

References

1.   Hunt SC, Hasstedt SJ, Kuida H, Stults BM, Hopkins PM, Williams RR
     Am. J. Epidemiol. 129, 625-638, 1989

2. Assmann G
   Lipid Metabolism and Atherosclerosis. Schattauer Verlag, Stuttgart 1982

3. Assmann G, Schulte H
   in: Lipid Metabolism Disorders and Coronary Heart Disease.
   MVV Medizin Verlag, München 1989, pp 87-121

4: Brown MS, Goldstein JL
   Science 232, 37-47, 1986

5. Lenzen HJ, Assmann G, Buchwalsky R, Schulte H
   Clin. Chem. 32, 778-781, 1986

6. Gregg RE, Zech LA, Schaefer EJ, Stark D, Wilson D, Brewer HBjr
   J. Clin. Invest. 78, 815-821, 1986

7. Wojciechowski AP, Farrall M, Cullen P, Wilson TME, Bayliss JD, Farren B, Griffin BA,
   Caslake MJ, Packard CJ, Shepherd J, Thakker R, Scott J
   Nature 349, 161-164, 1991

8. Pedersen JC, Berg K
   Clin. Genet. 35, 331-337, 1989

9. Seed BM, Hoppichler F, Reaveley D, McCarthy S, Thompson GR, Boerwinkle E, Utermann G
   New Engl. J. Med. 322, 1494-1499, 1990

10. Rees A, Shoulders CC, Stcks J, Galton D, Baralle FE
    Lancet I, 444-446, 1983

11. Ferns GAA, Stocks J, Ritchie C, Galton DJ
    Lancet II, 300-303, 1985

12. Ordovas JM, Schaefer EJ, Salem D
    N. Engl. J. Med. 314, 671-677, 1986

13. Coleman RT, Gonzales PA, Funke H, Assmann G, Levy-Wilson B, Frossard PM
    Mol. Biol. Med. 3, 213-228, 1986

14. Assmann G, Schmitz G, Funke H, von Eckardstein A
    Current Opin. Lipidol. 1, 110-115, 1990

15. Antonarakis SE, Oettgen P, Chakravarti A, Halloran SL, Hudson RR, Feisee L,
    Karathanasis SK
    Hum. Genet. 80, 265-273, 1988

16. von Eckardstein A, Funke H, Henke A, Altland K, Benninghoven A, Assmann G
    J. Clin. Invest. 84, 1722-1730, 1989

17. Funke H, von Eckardstein A, Pritchard PH, Albers JJ, Kastelein JJP, Droste C, Assmann G
    Proc. Natl. Acad. Sci. (USA), 1991, in press

18. Funke H, von Eckardstein A, Pritchard PH, Karas M, Albers JJ, Assmann G
    J. Clin. Invest. 87, 371-376, 1991

19. Steinmetz A, Assefbarkhi N, Eltze C, Ehlenz K, Funke H, Pies A, Assmann G, Kaffarnik H
    J. Lipid Res. 31, 1005-1013, 1990

20. Paulweber B, Wiebusch H, Miesenboeck G, Funke H, Assmann G, Hoelzl B, Sippl MJ, Friedl W, Patsch JR, Sandhofer F
    Atherosclerosis 86, 239-250, 1991

21 Funke H, von Eckardstein A, Pritchard PH, Hayden MR, Albers JJ, Jacotot B, Gerdes U, Assmann G
    Circulation 82, Suppl. III, 426, 1990 (abstract)

FACILITATIVE GLUCOSE TRANSPORTERS:

REGULATION AND POSSIBLE ROLE IN NIDDM

Mike Mueckler and Karen Tordjman

Department of Cell Biology and Physiology
Washington University School of Medicine
St. Louis, MO   63110

ABSTRACT

Glucose transport by facilitated diffusion is mediated by a family of tissue-specific membrane glycoproteins. Five members of this gene family have been identified by cDNA cloning, at least two of which (GLUT 1 and GLUT 4) are expressed in insulin-sensitive tissues. The GLUT 1 (HepG2-type) transporter is the more widely distributed of these two proteins. It provides many cells with their basal glucose requirement for ATP production and the biosynthesis of sugar-containing macromolecules. The GLUT 4 (adipocyte/muscle) transporter is expressed predominately in tissues that are insulin-sensitive with respect to glucose uptake. This protein is an excellent candidate for a highly specific genetic defect predisposing to insulin resistance.

We have studied the regulation of GLUT 1 and GLUT 4 expression in insulin-sensitive 3T3L1 adipocytes. The fibroblast-like preadipocytes express only the GLUT 1 isoform. Differentiation of the preadipocytes into adipocytes is accompanied by de novo expression of the GLUT 4 isoform and a relative decrease in the expression of GLUT 1. The increased expression of GLUT 4 is paralleled by the acquisition of insulin-sensitive glucose transport activity, whereas the decrease in GLUT 1 parallels the decrease in basal transport activity observed during the differentiation process. Insulin-stimulated glucose transport in the fully differentiated 3T3L1 adipocytes is due in part to the translocation of both GLUT 1 and GLUT 4 isoforms from an intracellular vesicular compartment to the plasma membrane. In addition to its acute effect, insulin exerts a chronic stimulation of transport activity by selectively increasing levels of GLUT 1 protein and mRNA. Tolbutamide has a similar chronic effect on GLUT 1 protein and mRNA and acts additively with insulin. Chronic glucose deprivation of these cells stimulates glucose uptake via the stimulation of GLUT 1 gene expression and translocation of GLUT 4 transporters to the plasma membrane.

DNA Polymorphisms as Disease Markers, Edited by D.J. Galton and
G. Asmann, Plenum Press, New York, 1991

INTRODUCTION

The transport of glucose across animal cell membranes is catalyzed by members of two distinct gene families [1]. Facilitative glucose transporters [2,3] are expressed ubiquitously in mammalian cells, whereas expression of the Na/glucose cotransporters [4] is restricted to epithelial cells of renal tubules and the intestinal mucosa. The Na-dependent proteins are secondary active transport systems that reside in the apical membranes of the epithelia and concentrate glucose in the cells from the intestinal contents and the forming urine. The facilitated diffusion transporters are passive systems that equilibrate sugar across membranes. These latter proteins are responsible for the movement of sugar from the blood into cells, supplying cellular glucose for energy metabolism and the biosynthesis of sugar-containing macromolecules such as glycoproteins, glycolipids, and nucleic acids. Additionally, facilitative glucose transport in certain tissues may play a critical role in organismal glucose homeostasis.

THE GLUT 1 AND GLUT 4 GLUCOSE TRANSPORTERS

Five facilitative glucose transporter isoforms have been identified by cDNA cloning (see Table 1). The human erythrocyte-type glucose transporter [5] was cloned from HepG2 cells in 1985 [6], and the equivalent rat isoform was subsequently cloned from brain [7]. These two cDNA species have been used to identify and clone novel transporters from liver/pancreatic islets [8-10], embryonic muscle [11], insulin-sensitive tissues (heart, fat, skeletal muscle) [12-15], and small intestine [3]. The functional identity of these proteins as transporters of glucose has been confirmed by expression studies [3,9,10,13,16-18]. The transporters are identified by the cell or tissue type from which they were first cloned, or based on a numbering system corresponding to the order in which they were identified and cloned.

Table 1. Mammalian Facilitative Glucose Transporters

Type	Tissue Distribution	Kinetic Properties	Regulatory Factors	Gene
HepG2	many tissues; abundant in brain, erythrocytes, placenta, immortal cell lines	human erythrocytes–asymmetric carrier with accelerated exchange; V_{max}(influx) < V_{max}(efflux); K_m ~ 5-30 mM (variable)	oncogenes, tumor promoters, growth factors, glucose deprivation, ATP, insulin, butyrate	GLUT 1
Liver	liver, pancreatic B cells, kidney, intestine (basolateral membrane)	liver–simple, symmetric carrier, K_m ~66 mM; intestine–asymmetric carrier, V_{max}(efflux) V_{max}(influx) K_m ~23-48 mM (variable)	?	GLUT 2
Fetal Muscle	many tissues; abundant in brain, kidney, placenta	?	?	GLUT 3
Fat/ Muscle	brown and white fat, red and white muscle, heart, smooth muscle(?)	Adipocyte–simple, symmetric carrier; K_m ~2.5–5 mM	insulin, excercise, beta-adrenergic agonists, streptozotocin-induced diabetes	GLUT 4
Small Intestine	small intestine, kidney, fat	?	?	GLUT 5

The HepG2 protein (GLUT 1 transporter) is the most widely distributed of the transporters. It is expressed in many adult and fetal tissues and is most abundant in placenta, brain (especially microvessels), and human erythrocytes [6,7,19-21]. This transporter appears to play primarily a "housekeeping" role, in that it appears to be involved in the survival of individual cells by providing them with their basal glucose requirement. The housekeeping role of GLUT 1 is supported by the observation that increased expression of the HepG2-type transporter occurs when cultured cells are starved for glucose [22,23]. Its expression is also induced by factors that stimulate cellular growth and division, such as oncogenes [24,25], polypeptide growth factors [26-29], and tumor promoters [24]. This response may be important in cells that are actively growing and dividing, and thus require increased levels of glucose for the biosynthesis of glycoproteins and nucleic acids and for the production of ATP. Thus, the HepG2 isoform is expressed prominently in most immortal cell lines.

The GLUT 4 transporter is the major isoform present in tissues that are insulin-sensitive with respect to glucose transport, i.e., brown and white adipose tissue, skeletal muscle, and heart [12,13,14,15]. Glucose transport is rate-limiting for its metabolism in these tissues [30,31], and skeletal muscle is the major site of glucose utilization in the postprandial state and in the presence of elevated insulin levels [32,33]. Thus, the regulation of transport in muscle plays a key role in glucose homeostasis. Insulin-stimulated transport in these tissues occurs, at least in part, via the redistribution of transporters from an intracellular membrane compartment to the cell surface [34-36]. The most obvious explanation for the expression of a common transporter in fat and muscle tissue is that this protein fulfills a unique role in the acute insulin-mediated increase in transport activity.

REGULATION OF GLUCOSE TRANSPORT IN INSULIN SENSITIVE CELLS

We have studied the regulation of glucose transport in murine 3T3L1 adipocytes. These cells grow and replicate in culture as fibroblast-like preadipocytes [37]. The confluent preadipocytes can be induced to differentiate into adipocyte-like cells in the presence of insulin, dexamethasone, and isobutyl methylxanthine. Unlike available muscle-derived cell lines, 3T3L1 adipocytes are exquisitely sensitive to insulin with respect to increased glucose transport. This may reflect the observation that 3T3L1 adipocytes express much higher levels of GLUT 4 than do muscle cell lines such as BC3H1 and L6. A cultured cell line offers the advantage over in vivo studies in that the direct effects of various agents on transport can be examined. However, one must be aware that an immortalized cell line may not behave identically to its ancestral cell of origin.

3T3L1 preadipocytes express only GLUT 1. No GLUT 4 mRNA or protein is detectable in the confluent fibroblast-like cultures (Fig. 1). During the differentiation process, GLUT 4 mRNA and protein accumulate as the cells round-up and acquire the adipocyte morphology. Differentiation is accompanied by a decrease in basal glucose transport, and an increase in the maximal acutely insulin-responsive glucose transport (Fig. 2). The decrease in basal transport parallels a decrease in the level of GLUT 1 expression, and the increase in acutely insulin-stimulated transport parallels the increased expression of GLUT 4. These data suggest that GLUT 1 is responsible for basal glucose transport in these cells, whereas GLUT 4 contributes only to insulin-responsive glucose transport.

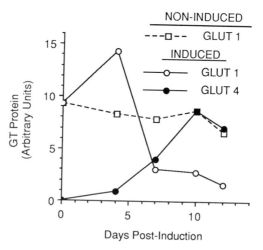

Fig. 1. GLUT 1 and GLUT 4 protein during 3T3L1 differentiation. Confluent cultures of 3T3L1 preadipocytes were induced to differentiate at day 0 by adding medium containing insulin, dexamethasone, and isobutylmethyl xanthine [29]. The differentiation medium was removed at day 2 and replaced with medium lacking the differentiation components for the remainder of the time course. The cells had acquired the adipocyte morphology by day six. Glucose transporter protein levels were quantitated in total detergent-soluble cellular extracts by immunoblotting using antisera specific for either GLUT 1 or GLUT 4. Data were taken from ref. 28.

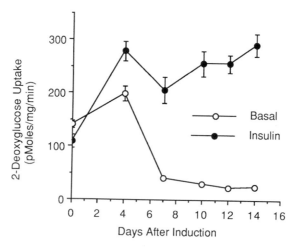

Fig. 2. Glucose transport activity during 3T3L1 differentiation. 3T3L1 preadipocytes were subjected to the differentiation protocol described in the legend to Fig. 1. 2-deoxyglucose uptake measurements were performed at room temperature for 6 min as described [29]. Insulin-treated cells were exposed to 100 nM insulin for 20 min prior to uptake measurements. Data from ref. 28.

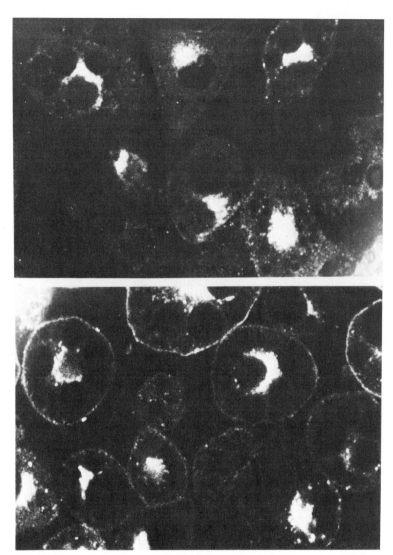

Fig. 3. Redistribution of GLUT 4 to the plasma membrane in response to acute insulin
treatment. 3T3L1 cells were grown and differentiated on glass coverslips and then
either treated with 100 nM insulin for 20 min (lower panel) or not (upper panel).
Cells were then fixed and subjected to indirect immunofluorescence staining using
primary antibodies specific for GLUT 4 and fluorescein isothiocyanate-conjugated
second antibody. Cells were then imaged using a laser confocal microscope.
Optical sections are shown. Plasma membrane staining is clearly distinguished
from intracellular staining in the lower panel by the halo effect. No plasma membrane staining is evident in untreated adipocytes.

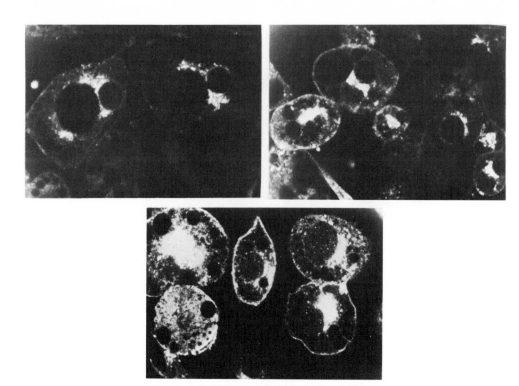

Fig. 4. Distribution of GLUT 1 in 3T3L1 adipocytes after acute and chronic insulin treatment. 3T3L1 adipocytes were left untreated (upper left panel) or treated with
100nM insulin for 20 min (upper right panel) or 3 days (lower panel). Cells were
subjected to indirect immunofluorescence using primary antibody specific for
GLUT 1 and fluorescein isothiocyanate-conjugated second antibody. Cells were
imaged using a laser confocal microscope. Note the weak, patchy plasma membrane staining in basal cells, as opposed to the total absence of GLUT 4 plasma
membrane staining in basal cells as shown in Fig. 3.

This hypothesis is supported by laser confocal microscopic images of adipocytes stained with antibodies specific for the GLUT 1 or GLUT 4 transporter
proteins. Only GLUT 1 is detectable in the plasma membrane of basal cells.
Insulin appears to increase the plasma membrane content of both GLUT 1 and
GLUT 4, although the bulk of both transporter isoforms is present in a dis-
crete, perinuclear, cytoplasmic compartment both in the presence and absence
of insulin (Fig. 3 and 4). The insulin-mediated redistribution of both transporter isoforms can be quantitated by immunoblot analysis of subcellular fractions
(not shown). In the experiments shown in Fig. 3 and 4, insulin brought about a
4-fold increase in the plasma membrane (PM) content of both GLUT 1 and
GLUT 4, with corresponding reductions in the transporter content of the low-
density microsomal (LDM) fraction. The LDM fraction contains the intracellular pool of glucose-transporter containing vesicles. The 4-fold increase in the
glucose transporter content of the PM fraction was accompanied by a 13-fold
increase in glucose transporter activity as assessed by 2-deoxyglucose uptake.

Because the transporter content of the PM fraction accurately reflects glucose transporters exposed at the cell surface [38], these data suggest that the intrinsic activity of one or both transporter isoforms must also increase in response to insulin in order to account fully for the total increase in transport function.

A major consequence of diabetes mellitus is a disruption in organismal glucose and insulin homeostasis. Chronic changes in blood glucose and insulin levels in diabetics may exacerbate insulin resistance by altering the expression of glucose transporter genes in peripheral tissues. In addition to the acute stimulatory effect of insulin on transport, both insulin and glucose exert direct, long-term effects on glucose transport in 3T3L1 adipocytes. Fig. 5 summarizes data concerning the chronic effects of insulin on glucose transport in 3T3L1 adipocytes. Treatment of adipocytes for 3 days results in a ~4-fold increase in 2-deoxyglucose uptake relative to untreated control cells, mediated by parallel increases in GLUT 1 protein and mRNA. No change was occurs in the steady-state levels of GLUT 4 mRNA or protein after chronic insulin administration. Thus, GLUT 1 appears to participate in both the acute and chronic phases of insulin-stimulated glucose transport, whereas GLUT 4 participates only in the acute phase. Interestingly, the oral hypoglycemic agent, tolbutamide, also increases glucose transport in 3T3L1 adipocytes via an increase in GLUT 1 mRNA and protein, without affecting GLUT 4 gene expression. Tolbutamide acts synergistically with insulin to increase 2-deoxyglucose uptake and GLUT 1 protein and mRNA levels.

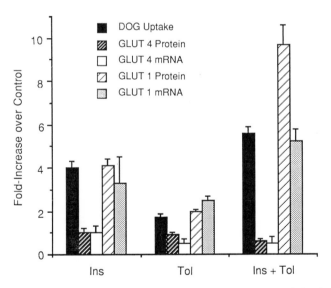

Fig. 5. Correlation between GLUT 1 gene expression and transport activity in 3T3L1 adipocytes after chronic treatment with insulin and/or tolbutamide. Adipocytes were left untreated (control) or were treated with 100 nM insulin, 1.5 mM tolbutamide, or both agents for 3 days. Media was changed daily. Steady-state mRNA and protein levels were quantitated by Northern and Western blot analyses as described [29]. Data from ref. 28.

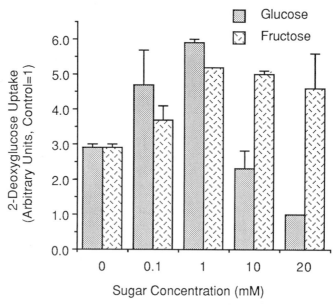

Fig. 6. Effect of glucose concentration on transport activity in 3T3L1 adipocytes. 2-deoxyglucose uptake measurements [29] were performed on adipocytes after 48 h incubation in medium containing the indicated concentration of glucose or fructose. Data from ref. 23.

The level of extracellular glucose exerts a direct effect on glucose transport in 3T3L1 adipocytes [23] (Fig. 6). Transport activity increases as the glucose concentration in the medium is decreased below 20 mM (the glucose concentration in which the cells are normally grown) and reaches a maximum between 0.1 and 1 mM glucose. The "repressive" effect of the high glucose concentration is specific to glucose and cannot be duplicated by substituting fructose for glucose in the medium. Glucose deprivation is accompanied by an increase in the level of GLUT 1 mRNA and a dramatic decrease in GLUT 4 mRNA (Fig. 7). These changes are understandable in terms of the physiologic roles proposed above for the GLUT 1 and GLUT 4 transporters. GLUT 1 protein in total cellular homogenates increases in parallel with but to an even greater extent than GLUT 1 mRNA. The total GLUT 1 protein continues to increase past the 24 h time point, at which transporter activity actually begins to decline (Fig. 8). However, much of this later increase in the protein is due to the accumulation of an aberrantly glycosylated lower molecular weight form of the GLUT 1 transporter (not shown). This lower molecular weight species is apparently mistargeted or nonfunctional and thus its accumulation does not contribute to increased transport activity. Curiously, despite the progressive disappearance of GLUT 4 mRNA during the course of glucose deprivation, the level of total cellular GLUT 4 protein does not change. Apparently this protein is selectively stabilized during glucose deprivation. Subcellular fractionation experiments suggest that the increased transport activity observed after 24 h of total glucose deprivation is due to the increased plasma membrane content of both GLUT 1 and GLUT 4 (not shown). Thus, glucose starvation increases transport activity in 3T3L1 adipocytes via a complicated series of events including increased synthesis of GLUT 1, decreased degradation of both GLUT 1 and GLUT 4 protein, and translocation of intracellular GLUT 4 molecules to the plasma membrane.

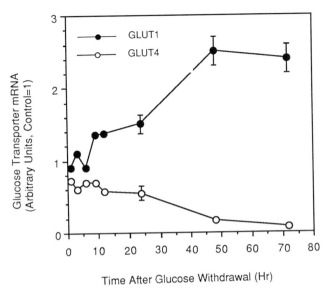

Fig. 7. GLUT 1 and GLUT 4 mRNA levels in 3T3L1 adipocytes during glucose starvation. Steady-state levels of glucose transporter mRNAs were determined by northern blotting at various time points after withdrawal of glucose. Data from ref. 23.

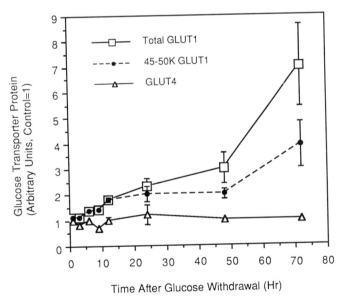

Fig. 8. GLUT 1 and GLUT 4 protein levels in 3T3L1 adipocytes during glucose starvation. Glucose transporter protein levels were determined by Western blot analysis at various time points after glucose withdrawal. Data from ref. 23.

DISCUSSION: GLUCOSE TRANSPORTERS AND NIDDM

Because of their role in the regulation of glucose homeostasis, the glucose transporters are obvious candidates for genetic defects predisposing to insulin-resistant states such as NIDDM. NIDDM is a heterogeneous genetic disease whose manifestation is influenced by environmental factors [39,40]. It is also likely to be polygenic in nature [41]. There are several ways in which glucose transporters could be directly involved in this disease. Specific alleles at one or more glucose transporter loci could predispose to NIDDM, and/or other genetic loci whose products regulate transporter activity, insulin-responsiveness, or gene expression, could be involved. In the following discussion, a "defect" in a glucose transporter refers to any of these possibilities.

Which glucose transporter is the most likely candidate for a defect predisposing to NIDDM? The HepG2 transporter is widely distributed in adult and fetal tissues and appears to be a major transporter expressed in the mammalian brain. Genetic defects that result in severe disruption of this protein or its expression would probably be lethal <u>in utero,</u> and more mild defects would most likely result in non-diabetic phenotypes. Although it is possible that a decrease in the activity of this protein or its expression could affect whole body glucose disposal, it is unlikely to be directly involved in the pathogenesis of NIDDM because of its tissue distribution.

The liver/islet (GLUT 2) transporter is interesting in that a defect in this single gene product could hypothetically produce the two principal features of NIDDM– relative hypoinsulinemia and insulin resistance (see Fig. 9). Although transport is not normally rate-limiting for glucose uptake into either islets or hepatocytes, a defect in this transporter could result in reduced insulin biosynthesis and secretion, as well as reduced uptake and metabolism of glucose in liver.

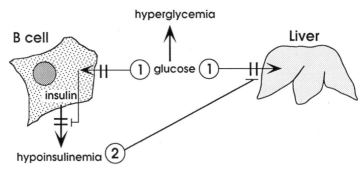

Fig. 9. Hypothetical role of the liver/islet glucose transporter in the pathogenesis of NIDDM. (1) A primary genetic defect in the liver/islet glucose transporter could give rise directly to both insulin resistance at the level of the liver, resulting in postprandial hyperglycemia, and relative hypoinsulinemia due to diminished uptake and metabolism of glucose in pancreatic B cells. (2) Protracted hypoinsulinemia could further exacerbate insulin resistance at the liver, e.g., by decreasing expression of hepatic glucokinase. Glucose transport into these tissues is not normally rate-limiting for its metabolism, and thus the defect would have to be of sufficient magnitude to noticeably reduce the steady-state intracellular concentration of free glucose, in order to affect the rate of the phosphorylation reaction and subsequent metabolism. There appears to be no evidence for a defect in hepatic glucose uptake in NIDDM [42,43]. Taken from ref. 2.

This result would follow if the defect reduced the velocity of transport to a level that diminished the steady-state concentration of intracellular glucose, thus affecting the velocity of the glucokinase reaction and the overall rate of glucose metabolism in the β cell and the hepatocyte. Such a defect might be reasonably specific for a diabetic phenotype because of the limited tissue distribution of the GLUT 2 transporter. However, there are at least two problems with this hypothesis. First, the fraction of whole body glucose disposal that occurs via the splanchnic bed is insufficient to account quantitatively for insulin resistance in NIDDM . Second, the available data indicate that hepatic glucose uptake is not abnormal in type II diabetics [42]. Thus, if the liver/islet transporter is involved in the pathogenesis of NIDDM, its effect is most likely at the level of the β-cell.

The major site of insulin resistance in NIDDM is skeletal muscle [42,43,44, [45]. Because GLUT 4 is the major insulin-responsive glucose transporter in this tissue, it is an excellent hypothetical target for a defect giving rise to insulin resistance (see Fig. 10). The following points should be considered in this regard: 1) Transport of glucose appears to be rate-limiting for its utilization in muscle and therefore any reduction in the rate of transport would give rise to a proportional decrease in muscle glucose disposal; 2) The expression of the GLUT 4 isoform appears to be restricted to insulin-sensitive tissues and GLUT 4 is the isoform directly responsible for insulin-sensitive transport. The only predicted consequence of a genetic defect in this protein is insulin resistance; 3) Several recent studies [46-50] have shown a decrease in GLUT 4 levels in adipose and muscle tissue obtained from streptozotocin-treated, diabetic rats. However, reductions in GLUT 4 protein have not been detected in skeletal muscle obtained from human NIDDM subjects [51]; and 4) The expected phenotypic progression of a genetic defect in this transporter is similar to that which can be inferred from a revealing study of NIDDM subjects and their first degree relatives [52] (see Fig. 10).

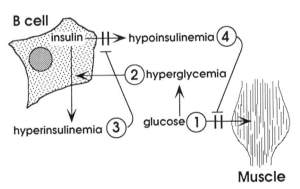

Fig. 10. Hypothetical role of the adipocyte/muscle glucose transporter in the pathogenesis of NIDDM. (1) A primary genetic defect in the adipocyte/muscle glucose transporter would be highly specific for an insulin-resistant phenotype at the level of muscle and fat. (2) The resulting postprandial hyperglycemia would initially lead to transient bouts of hyperinsulinemia. (3) Continued over-stimulation of B cells could, in susceptible individuals, result in eventual B cell damage and consequent reduced synthesis and secretion of insulin. (4) The resulting absolute hypoinsulinemia might further exacerbate insulin resistance in muscle, e.g., by reducing expression of a glucose transporter gene in this tissue. This scenario is reasonably consistent with clinical data on NIDDM subjects and their first-degree relatives [52,53]. Taken from ref. 2.

At present there is no compelling evidence for the direct involvement of glucose transporter gene defects in NIDDM. One study has been published [54] indicating an association between restriction fragment length polymorphisms at the HepG2 glucose transporter locus (GLUT [55]) and NIDDM, but no association was observed in two other studies [56,57]. However, the interpretation of these studies is a formidable problem because of the difficulty involved in the analysis of complex human genetic traits [58]. If NIDDM is both polygenic and heterogeneous, population studies may be of little practical value. The relatively late age of onset of diseases such as NIDDM makes it very difficult to acquire good pedigrees for family studies, and sophisticated genetic analyses using the human linkage map will undoubtedly be required [59].

Irrespective of whether glucose transporter genes are directly involved in the predisposition to NIDDM, understanding the regulation of glucose transport in insulin-sensitive tissues is likely to be an integral part of unraveling the molecular basis of insulin resistance, and perhaps identifying additional candidate genes for NIDDM. The observation that NIDDM does not result in a reduction in GLUT 4 levels in skeletal muscle rules out the simple possibility that decreased expression of this protein is responsible for the insulin-resistance accompanying this disease. Additional possibilities that must be explored include defects in transporter intrinsic activity, in the ability of the transporter to redistribute to the plasma membrane in response to insulin, and in the insulin receptor and its signaling mechanism.

ACKNOWLEDGMENTS

Work in the authors' laboratory is supported by research grants from the National Institutes of Health, the Juvenile Diabetes Foundation, and the American Heart Association. M. M. is a recipient of a Career Development Award from the Juvenile Diabetes Foundation.

REFERENCES

[1] Baldwin SA, Henderson PJF: Homologies between sugar transporters from eukaryotes and procaryotes. Annu Rev Physiol 51:459-71, 1989

[2] Mueckler M, Family of glucose transporter genes: Implications for glucose homeostasis and diabetes, Diabetes 39:6-11, 1990.

[3] Bell GI, Kayano, T, Buse JB, Burant, CF, Takeda J, Lin D, Fukumoto H, Seino S, Diabetes Care 13:198-208, 1990.

[4] Wright JK, Seckler, R, Overath, P: Molecular aspects of sugar ion cotransport. Annu Re. Biochem 55:225-48, 1986

[5] Mueckler M: Structure and function of the glucose transporter. In Red Blood Cell Membranes. Agre P and Parker JC, Eds. New York, Marcel Dekker, Inc., 1989, p. 31-45.

[6] Mueckler M, Caruso C, Baldwin S, Panico M, Blench I, Morris HR, Allard W, Lienhard GE, Lodish HF: Sequence and structure of a human glucose transporter. Science 229:941-45, 1985

[7] Birnbaum MJ, Haspel HC, Rosen OM: Cloning and characterization of a cDNA encoding the rat brain glucose-transporting protein. Proc Natl Acad Sci USA 83:5784-88

[8] Fukumoto H, Seino S, Imura H, Seino Y, Eddy RL, Fukushima Y, Byers MG, Shows TB, Bell GI: Sequence, tissue distribution, and chromosomal localization of mRNA encoding a human glucose transporter-like protein. Proc Natl Acad Sci USA 85:5434-38, 1988

[9] Thorens B, Sarkar HK, Kaback HR, Lodish HF: Cloning and functional expression in bacteria of a novel glucose transporter present in liver, intestine, kidney, and β-pancreatic islet cells. Cell 55:281-90, 1988

[10] Permutt MA, Koranyi L, Keller K, Lacy P, Sharp D, Mueckler M: Molecular cloning and functional expression of a human pancreatic islet glucose transporter cDNA. Proc Natl Acad Sci USA, in press.

[11] Kayano T, Fukumoto H, Eddy RL, Fan Y, Byers MG, Shows TN, Bell GI: Evidence for a family of human glucose transporter-like proteins. J Biol Chem 263:15245-48, 1988

[12] James DE, Strube M, Mueckler M: Molecular cloning and characterization of an insulin-regulatable glucose transporter. Nature 33: 83-87, 1989

[13] Birnbaum MJ: Identification of a novel gene encoding an insulin-responsive glucose transporter protein. Cell 57:305-15, 1989

[14] Charron MJ, Brosius FC, Alper SL, Lodish HF: A glucose transport protein expressed predominately in insulin-responsive tissues. Proc Natl Acad Sci USA 86:2535-39, 1989

[15] Fukumoto H, Kayano, T, Buse JB, Edwards Y, Pilch PF, Bell GI, Seino S: Cloning and characterization of the major insulin-responsive glucose transporter expressed in human skeletal muscle and other insulin-responsive tissues. J Biol Chem 264:7776-79, 1989

[16] Keller K, Strube M, Mueckler M: Functional expression of the human HepG2 and rat adipocyte glucose transporters in Xenopus oocytes. Comparison of kinetic parameters. J Biol Chem 264: 18884-9, 1989

[17] Vera JC, Rosen OM: Functional expression of mammalian glucose transporters in Xenopus laevis oocytes: Evidence for cell-dependent insulin sensitivity. Mol Cell Biol 9:4187-95, 1989

[18] Gould GW, Lienhard GE: Expression of a functional glucose transporter in Xenopus oocytes. Biochemistry 28: 9447-52, 1989

[19] Flier JS, Mueckler M, McCall AL, Lodish HF: Distribution of glucose transporter messenger RNA transcripts in tissues of rat and man. J Clin Invest 79:657-61, 1987

[20] Fukumoto H, Seino S, Imura H, Seino Y, Bell GI: Characterization and expression of human HepG2/erythrocyte glucose transporter gene. Diabetes 37:657-61, 1988

[21] Sadiq F, Holtzclaw L, Chundu, K, Muzzafar, A, and Devaskar, S: The ontogeny of the rabbit brain glucose transporter. Endocrinology 126: 2417-24, 1990

[22] Haspel HC, Wilk EW, Birnbaum MJ, Cushman SW, Rosen, OM: Glucose deprivation and hexose transporter polypeptides of murine fibroblasts. J Biol Chem 261:6778-89, 1986

[23] Tordjman KM, Leingang K, Mueckler M: Differential regulation of the HepG2 and adipocyte/muscle glucose transporters in 3T3L1 adipocytes. Effect of chronic glucose deprivation. Biochem J 271: 201-207, 1990

[24] Flier JS, Mueckler MM, Usher P, Lodish HF: Elevated levels of glucose transport and transporter messenger RNA are induced by ras or src oncogenes. Science 235:1492-95, 1987

[25] Birnbaum MJ, Haspel HC, Rosen OM: Transformation of rat fibroblasts by FSV rapidly increases glucsose transporter gene transcription. Science 235:1495-98, 1987

[26] Hiraki Y, Rosen OM, Birnbaum MJ: Growth factors rapidly induce ex-

pression of the glucose transporter gene. J Biol Chem 263:13655-62, 1988

[27] de Herreros AG, Birnbaum MJ: The regulation by insulin of glucose transporter gene expression in 3T3 adipocytes. J Biol Chem 264:9885-90, 1989

[28] Rollins BJ, Morrison, ED, Usher P, Flier JS: Platelet-derived growth factor regulates glucose transporter gene expression. J Biol Chem 263:16523-26, 1988

[29] Tordjman K, Leingang, K, James, DE, Mueckler M: Differential regulation of two distinct glucose transporter species expressed in 3T3L1 adipocytes. Effect of chronic insulin and tolbutamide treatment. Proc Natl Acad Sci USA 86: 7761-5

[30] Crofford OB, Renold AE: Glucose uptake by incubated rat epididymal adipose tissue. Rate-limiting steps and site of insulin action. J Biol Chem 240:3237-44, 1965

[31] Berger M, Hagg S, Ruderman NB: Glucose metabolism in perfused skeletal muscle. Interaction of insulin and exercise on glucose uptake. Biochem. J 146:231-38, 1975

[32] Ferrannini E, Bjorkman O, Reichard G, Pilo A, Olsson M, Wahren J, DeFronzo RA: The disposal of an oral glucose load in healthy subjects. Diabetes 34:580-88, 1985

[33] James DE, Jenkins AB, Kraegen EW: Heterogeneity of insulin action in individual muscles in vivo:euglycemic clamp studies in rats. Am J Physiol 248:E575-E580, 1985

[34] Suzuki I, Kono T: Evidence that insulin causes translocation of glucose transport activity to the plasma membrane from an intracellular storage site. Proc. Natl Acad Sci USA 77:2542-45, 1980

[35] Cushman SW, Wardzala LJ: Potential mechanism of insulin action on glucose transport in the isolated rat adipose cell. Apparent translcoation of intracellular transport systems to the plasma membrane. J Bio. Chem 255:4758-62, 1980

[36] Wardzala LJ, Jeanrenaud B: Potential mechanism of insulin action on glucose transport in the isolated rat diaphragm. J Biol Chem 253:8002-5, 1981

[37] Green H, Kehinde O: Sublines of mouse 3T3 cells that accumulate lipid. Cell 1: 113-116, 1974

[38] Gould GW, Derechin V, James DE, Tordjman K, Ahern S, Gibbs EM, Lienhard GE, Mueckler M: Insulin-stimulated translocation of the HepG2/erythrocyte type glucose transporter expressed in 3T3-L1 adipocytes. J Biol Chem 264: 2180-4, 1989

[39] Rotter JI, Rimoin DL: Heterogeneity in diabetes mellitus–update, 1978. Diabetes 27:599, 1978

[40] Barnett AH, Eff C, Leslie RD, Pyke DA. Diabetes in identical twins: a study of 200 pairs. Diabetologia 20:87-93, 1981

[41] Friedman JM, Fialkow PJ: The genetics of diabetes. In Progress in Medical Genetics IV, Steinberg AG, Bearn AG, Motulsky AG, Childs B, Eds., Philadelphia, WB Saunders Co, p 199, 1980

[42] DeFronzo RA, Gunnarsson R, Bjorkman O, Olsson M, Wahren, J: Effects of insulin on peripheral and splanchic glucose metabolism in noninsulin dependent (type II) diabetes mellitus. J Clin Invest 76:149-55, 1985

[43] Ferrannini E, Simonson DC, Katz LD, Reichard G, Bevilacqua S, Barrett

EF, Olsson M, DeFronzo RA: The disposal of an oral glucose load in patients with non-insulin dependent diabetes. Metabolism 37:79-85, 1988

[44] Firth RG, Bill PM, Marsh HM, Hansen I, Rizza RA: Postprandial hyperglycemia in patients with noninsulin-dependent diabetes mellitus. J Clin Invest 77:1525-32, 1986

[45] Shulman GI, Rothman DL, Jue T, Stein P, DeFronzo RA, Shulman RG: Quantitation of muscle glycogen synthesis in normal subjects and subjects with non-insulin-dependent diabetes by 13-C nuclear magnetic resonance spectroscopy. N Engl J Med 322: 223-28, 1990

[46] Garvey WT, Huecksteadt TP, Birnbaum MJ: Pretranslational suppression of an insulin-responsive glucose transporter in rats with diabetes mellitus. Science 245: 60-63, 1989

[47] Berger, J, Biswas, C, Vicario, PP, Strout HV, Saperstein R, Pilch, PF: Decreased expression of the insulin-responsive glucose transporter in diabetes and fasting. Nature 340:70-74, 1989

[48] Sivitz WI, DeSautel SL, Kayano T, Bell GI, Pessin JE: Regulation of glucose transporter messenger RNA in insulin-deficient states. Nature 340:72-74, 1989

[49] Garvey WT, Huecksteadt TP, Matthaei S, Olefsky JM: The role of glucose transporters in the cellular insulin resistance of type II noninsulin-dependent diabetes mellitus. J Clin Invest 81:1528-36, 1988

[50] Bourey RE, Koranyi L, James DE, Mueckler M, Permutt MA: Effects of altered glucose homeostasis on glucose transporter expression in skeletal muscle of the rat. J Clin Invest 86: 542-47, 1990

[51] Pedersen O, Bak JF, Andersen PH, Lund S, Moller DE, Flier JS, Kahn BB: Evidence against altered expression of GLUT 1 or GLUT 4 in skeletal muscle of patients with obesity or NIDDM. Diabetes 39: 865-70, 1990

[52] Eriksson J, Franssila-Kalluni A, Ekstrand A, Saloranta C, Widen E, Schalin C, Groop L. Early metabolic defects in persons at increased risk for non-insulin-dependent diabetes mellitus. N Eng J Med 321:337-343, 1989

[53] DeFronzo, RA: The Triumvirate: β-cell, muscle, liver. A collusion responsible for NIDDM. Diabetes 37:667-87, 1988

[54] Li SR, Oelbaum RS, Baroni MG, Stock J, Galton DJ: Association of genetic variant of the glucose transporter with non-insulin-dependent diabetes mellitus. The Lancet Aug. 13:368-70, 1988

[55] Shows TB, Eddy RL, Byers MG, Fukushima Y, Dehaven CR, Murray JC, Bell GI: Polymorphic human glucose transporter gene (GLUT) is on chromosome 1p31.3→p35. Diabetes 36:546-49, 1987

[56] Xiang K-S, Cox NJ, Sanz N, Huang P, Karam JH, Bell GI: Insulin-receptor and apolipoprotein genes contribute to development of NIDDM in Chinese Americans. Diabetes 38:17-23, 1989

[57] Kaku K, Matsutani A, Mueckler M, Permutt MA: Polymorphisms of the HepG2/erythrocyte glucose transporter gene and non-insulin dependent diabetes mellitus. Diabetes 39: 49-56, 1990

[58] Cox NJ, Bell GI: Disease associations. Chance, artifact, or susceptibility genes? Diabetes 38:947-50, 1989

[59] Lander ES, Botstein D: Strategies for studying heterogeneous traits in humans by using a linkage map of restriction fragment length polymorphisms. Proc Natl Acad Sci USA 83:7353-57, 1986

MOLECULAR GENETIC APPROACH TO POLYGENIC DISEASE:
NON-INSULIN DEPENDENT DIABETES MELLITUS (NIDDM)

M. Alan Permutt, Laszlo Koranyi, and Akira Matsutani

Metabolism Division
Washington University School of Medicine
St. Louis, Missouri

I. INTRODUCTION

Diabetes is a complex metabolic disorder of carbohydrate, lipid, and protein metabolism which is secondary to insulin deficiency. The insulin deficiency varies in degree, being complete in insulin dependent diabetes (IDDM), and relative in non-insulin dependent diabetes (NIDDM). The strong association of certain HLA immune-response genes on chromosome 6 with IDDM provides a prominent genetic marker to distinguish these two types of diabetes (Trucco and Dormon, 1989). This genetic marker serves to aid in genetic counseling and provides insight into the etiology of IDDM.

In contrast to IDDM, there have been no genetic markers defined for NIDDM. The search for these markers is indicated by evidence of the strong genetic component of the disease (Rotter et al, 1990). The best evidence for genetic susceptibility to NIDDM is provided by concordance studies in twins (Rotter et al, 1990). Concordance for NIDDM in sibs and non-identical twins has been less than 40%, while in identical twins it has been shown to be greater than 90%. Environmental factors such as diet and obesity have marked effects on the incidence, yet prominent differences have been noted in the prevalence of NIDDM between countries, between racial groups in the same country, and between racial groups undergoing migration (see Zimmet, 1982, for review). For example, the prevalence of NIDDM in the United States is 2-4% for Caucasians, 4-6% for American Blacks, 10-15% for Mexican-Americans, and 35% for Pima Indians (Zimmet, 1982; Harris et al, 1987). While almost all Pimas Indians are obese, non-diabetic Pima Indians exhibit insulin resistance compared to age and weight matched non-diabetic Caucasians (Nagulesparan et al, 1982). Furthermore, the risk of diabetes has been shown to be greater for a Pima with two diabetic parents as compared to one or zero (Knowler et al, 1981), and the resistance to glucose utilization following maximal insulin stimulation was shown to be at least in part a familial trait, independent of the degree of obesity, age, sex, or physical fitness.

II. METHODS OF GENETIC ANALYSIS

Evaluating Candidate Gene Loci

The genetic factors which contribute to susceptibility to NIDDM can now be addressed with molecular genetic techniques. There are two general methods available to assess genetic components of a disease. The first uses random genetic markers, scattered throughout the genome, to "map" in families a particular phenotype to a region of chromo-

DNA Polymorphisms as Disease Markers, Edited by D.J. Galton and
G. Asmann, Plenum Press, New York, 1991

somal DNA, followed by molecular techniques to identify the defective gene. This process has been very effective in identifying single major gene defects as for example in cystic fibrosis (Barker, 1990). Another method to assess genes which might be defective in a complex metabolic disorder such as diabetes has been the so called "candidate gene" approach (Lusis, 1988). Here one chooses a locus likely to be involved and investigates the occurrence of abnormalities in this gene in affected and unaffected individuals, either through linkage analysis in families, or studies of populations of unrelated individuals.

Unlike the codominant inheritance of a single gene defect resulting in the cystic fibrosis phenotype for example, it is likely that evaluation of candidate loci for NIDDM will be more difficult. Common diabetic phenotypes, characterized by impaired insulin production and/or insulin action, might be polygenic (i.e. the combined result of several defects). If diabetic phenotypes are the result of many different gene defects, it might be possible to subclassify the phenotypes to better assess the genes involved. Subphenotypes have been used to study other polygenic diseases such as atherosclerosis (Antonarakis et al, 1988) and familial hypercholesterolemia (Lehrman et al, 1987). At the present time there has been little effort to divide individuals with NIDDM into subpopulations, although recently individuals within families have been characterized by metabolic studies as to impairment of insulin action (Eriksson et al, 1989), and these subphenotypic characterizations should prove useful in genetic analysis.

Animal Models for Genetic Analysis: Relevance to NIDDM in Man

Animal models have several important advantages for genetic analysis of diabetes susceptibility (Coleman and Hummel, 1975; Leiter et al, 1987): 1) environmental factors which can influence insulin production and action can be better controlled, 2) detailed analyses of tissues such as brain, skeletal muscle, and pancreas are not possible in man, and 3) genetic analysis in man is limited to available individuals. For example, the diabetes (db) mutation on chromosome 4 in the homozygous state results in spontaneous obesity in mice (Leiter et al, 1987; Kaku et al, 1989). The initial adaptation to the insulin resistance of the obesity is one of islet hyperplasia, with marked hyperinsulinemia in all inbred strains of mice. Some strains, however, develope pancreatic islet beta-cell necrosis and severe insulinopenic diabetes. The response to the obesity-induced insulin resistance in other inbred strains of mice is a milder diabetes, more like that of NIDDM in man. Genetic analysis in inbred of strains of mice has indicated that the obesity-induced diabetes phenotype is determined by multiple genes (Coleman and Hummel, 1975; Leiter et al, 1987; Kaku et al, 1989). These results in inbred mouse strains have obvious implications for NIDDM in man.

III. CANDIDATE LOCI

A. The Insulin Gene

Structural gene muatations, associated with either failure of conversion of proinsulin to insulin, or a biologically defective insulin resulting in various degrees of hyperinsulinemia, have been observed (Haneda et al, 1984). The majority of individuals with NIDDM have either relative or absolute hypoinsulinemia, however, which would more likely result from transcriptional mutations, e.g. nonsense mutations, or mutations in the promoter region, which would result in diminished insulin production. These mutations would not be detected at the protein level. Thus DNA polymorphisms at the insulin locus have been used to evaluate this possibility.

The 1430 base pair insulin gene (Figure 1) is composed of three exons separated by introns of 179 base pairs and 786 base pairs of DNA respectively (Ullrich et al, 1980). The gene has been mapped (Owerbach et al, 1980) to the telomeric region of the short arm of chromosome 11 (11p15), adjacent to the human insulin-like growth factor II (IGF-II) and tyrosine hydroxylase (TH) genes. Thus the locus spans approximately 45 kb of genomic DNA on chromosome ll. These genes exhibit different tissue specific expression and thus do not appear to be coordinately regulated. Approximately 375 base pairs upstream from the transcription initiation site of the insulin gene is a hypervariable region of DNA, comprised of variable numbers of tandem repeats (VNTR) (Bell et al, 1981). Several differences in the composition of the repeat units in various individuals also occurs (Bell et al, 1982). From 30-139 repeats of the 14-15 base pair unit are commonly seen, while in some individuals as many as 540 repeats have been observed. Thus the size variation is quite large, and the exact number of alleles in any racial group has not been defined. At a similar hypervariable region in the apoB gene, as many as 14 alleles were defined by use of the polymerase chain reaction and improved electrophoretic resolution of repeat units (Boerwinkle et al, 1989).

Based on the apparent non-random distribution of sizes within racial groups (Bell et al, 1981; Bell et al, 1982), the hypervariable region of the insulin gene has been divided into three classes. Alleles with approximately 40 repeat units (570 base pairs) were called class 1, larger alleles with 170 copies of the repeat (2470 base pairs) class 3, and intermediate sized alleles with approximately 95 copies of repeats class 2. For most populations these alleles fall into a bimodal distribution around class 1 and class 3, while in African and American Black populations the intermediate size (class 2) alleles are also quite common (Elbein et al, 1985). The extent of polymorphism varies markedly in racial groups, and the degree of heterogenity greater than that indicated by this classification. (Permutt and Elbein, 1990).

A number of DNA site polymorphisms have been identified (Elbein et al, 1988a; Elbein et al, 1985; Cox et al, 1988b; Xiang et al, 1987; Xiang and Bell, 1988) at the insulin gene locus (Figure 1), which used in conjunction with the hypervariable alleles makes this locus highly informative for linkage analysis in families, with heterozygosity greater than 90%.

Associations of Insulin Gene Polymorphisms with Diabetes

A positive association of the large class 3 allele at the VNTR region of the insulin gene with NIDDM was noted by a number of investigators (see Permutt and Elbein, 1990, for review). Following these initial reports, larger racially uniform populations were studied for associations, and these are summarized in Table 1. In Caucasians, one study found an association of class 1 alleles with NIDDM, while another found an association of class 3 alleles with NIDDM (Hitman et al, 1985). No associations of insulin VNTR polymorphic alleles with NIDDM were found in seven other racial groups (Elbein et al, 1985; Kanazawa et al, 1988; Frazier et al, 1986; Knowler et al, 1984; Haneda et al, 1986; Awata et al, 1985; Hitman et al, 1985; Aoyama et al, 1986; Takeda et al, 1986; Xiang et al, 1989). Associations of the smaller class 1 allele with IDDM were found in Caucasians, and in a pooled data set from several institutions this association has been confirmed, despite lack of sharing of these alleles among sib pairs with IDDM (Spielman et al, 1989; Cox et al, 1988; Cox and Spielman, 1989).

Linkage Analysis in Families

Lack of association between the insulin gene and NIDDM does not eliminate it from consideration (see Summary and Analysis). Linkage analysis within families represents another test of the involvement of a candidate gene locus. The autosomal dominant subtype of NIDDM or maturity onset diabetes of the young (MODY),includes some pedigrees with hypoinsulinemic individuals, and insulin gene mutations have been considered (Fajans SS,1990). Studies of several of these pedigrees have demonstrated lack of linkage, suggesting that this locus is not involved in the pathogenesis of MODY (Permutt and Elbein, 1990). Linkage analysis in more typical NIDDM families was reported in Utah Mormons, representative of Northern European Caucasians (Elbein et al, 1988b). Twelve families were informative for linkage, and lack of linkage was convincingly demonstrated in six. The conclusion of this analysis was that there was no linkage of the insulin gene and NIDDM under several models, and that insulin gene defects are not a major factor in the predisposition to NIDDM in Northern European Caucasians. However if the disease were polygenic, this interpretation may not be correct, although sib pair analysis should have shown a trend toward sharing. A similar conclusion was reached by Cox et al, who examined 20 American Black pedigrees (Cox et al, 1989). Lack of linkage in several other Caucasian pedigrees has also been demonstrated (Elbein et al, 1988b).

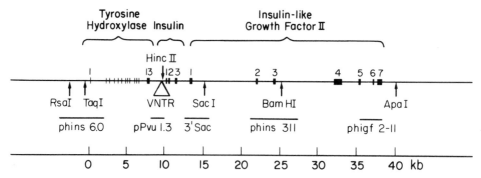

Fig. 1. A composite map of the human tyrosine hydroxylase-insulin-insulin-like growth factor II (TH-INS-IGF II) gene on chromosome 11p, adapted from (O'Malley and Rotwein 1988; Elbein et al, 1985a; Cox, Bell, Xiang 1988). The vertical lines and boxes represent the approximate sizes of the respective exons, numbered for each gene. The vertical arrows refer to polymorphic sites, and VNTR (variable number of tandem repeats) is the hypervariable region. The Rsa I, Taq I, Hinc II, Sac I, and VNTR polymorphisms and their respective genomic probes (horizontal lines) are described (Elbein et al, 1985a), and the Bam HI and Apa I polymorphisms and probes described by Xiang et al (1987, 1988). (Figure printed with permission of Balliere Tindall Limited, London, England, 1990).

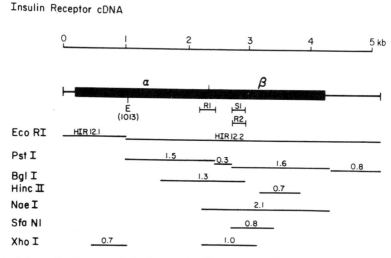

Fig. 2. A. Schematic diagram of the human insulin receptor cDNA. The untranslated 5' and 3' regions are represented by the thin horizontal line, and the translated region by the boxes. The regions encoding the alpha and beta subunits, the Eco RI site (E) at 1013, and the approximate locations of the two Rsa I (R1,R2) and two Sst I (S1,S2) RFLPs are indicated. The cDNA probes used by various labs to study the Rsa and Sst RFLPs, as well as other RFLPs, are restriction endo-nuclease fragments of the original Eco RI clones as indicated. The numbers refer to the sizes of the probes in kb. **B.** Schematic diagram of the human insulin receptor gene on chromosome 19 as adapted from Seino et al (1990). The exons, numbered 1-22, are indicated by boxes. The prorecceptor processing region is in exon 12. The R_1, R_2, B1 RFLPs have been mapped to the regions indicated, and the S1 RFLP between exons 14-15 (Elbein, 1989). The other RFLPs have not been mapped. (Figure printed with permission of Balliere Tindall Limited, London.)

Table 1. Associations of the Polymorphic VNTR Region of the Insulin Gene with NIDDM

Population	Control (n)	NIDDM	Association	Reference
Nauruan	56	85	N.S.	Serjeantson et al 1983
Japanese	64	51	N.S.	Awata et al 1983
Caucasian	83	76	p=0.025 (class 1)	Bell et al 1984
Caucasian	88	71	p<0.01 (class 3)	Hitman et al 1984
Pima Indian	38	49	N.S.	Knowler et al 1984
American Black	264	308	N.S.	Elbein et al 1985
Japanese	55	77	N.S.	Aoyama et al 1985
Japanese	47	52	N.S.	Haneda et al 1986
Japanese	49	68	N.S.	Takeda et al 1986
Mexican-American	36	37	N.S.	Frazier et al 1986
Tunisians	46	45	N.S.	Frazier et al 1986
Japanese	270	326	N.S.	Kanazawa et al 1988
Chinese Americans	146	186	N.S.	Xiang et al 1988

(Printed with permission of Balliere Tindall Limited, London.)

B. Human Insulin Receptor Gene

A prominent phenotypic characteristics of NIDDM is impaired insulin action (DeFronzo, 1988). Because all insulin action is mediated through its receptor, this protein has been considered a major candidate locus which might contribute to the genetic susceptibility to NIDDM (Permutt, 1990, Taylor,et al, 1990). Since the gene for the receptor was isolated in 1985, many studies of patients with isolated mutations, populations, and families have been reported.

The approximately 5 kb human insulin receptor cDNA (Figure 2) is encoded by 22 exons which extend over a region of genomic DNA of >120,000 base pairs on the short arm of chromosome 19 (Ebina et al, 1985; Ullrich et al, 1985). The cDNA sequence predicts a 1382 amino acid precursor, the alpha subunit comprising the amino terminal insulin binding portion of the precursor, and the beta subunit containing a transmembrane domain and the tyrosine-specific protein kinase portion. Nine different mutations of the receptor gene have been described in patients associated with insulin resistance and diabetes (see Seino et al, 1990, and Taylor et al, 1990, for reviews).

The Insulin Receptor Gene and NIDDM

Studies of mutant insulin receptor genes revealed several important features which may be relevant to the insulin resistance of NIDDM. First, 9 different mutant alleles were observed in 8 patients, and thus we can anticipate genetic heterogeneity within this locus. Second the mutations exhibited pleiotropic effects with respect to function, as obligate heterozygotes varied from normal to mild insulin resistance. Because the insulin receptor is a heterotetramer composed of 2 alpha and 2 beta subunits, one would predict that a single defective allele in an individual would result in perhaps only 1/4 the normal number of receptors, provided there was no compensation. The functional consequences of each mutation, however, must be determined.

Few studies have yet to determine how common the mutations described may be in NIDDM subjects. Taylor's laboratory screened 160 normal alleles for the phenylalanine 382 to valine mutation and found none (Accili et al, 1989). One hundred and three Japanese normal individuals were screened for a deletion of exons 17-22, and none were found (Taira et al, 1989). While these methods may be used to screen insulin receptor genes in the NIDDM population, even ifinsulin receptor mutations are common, the heterogeneity described in these genetic syndromes suggests there would be considerable heterogeneity within the NIDDM population.

RFLPs at the Insulin Receptor Locus and Population Studies

Evidence from population and family studies that defects in the insulin receptor gene occur in a substantial number of individuals with NIDDM has been reviewed(Permutt, 1990, Taylor et al, 1990). The insulin receptor gene extends over at least 120,000 base pairs of genomic DNA (Seino et al, 1990), and 23 different RFLPs have been described (see Taylor et al, 1990, for review). These polymorphisms have been defined in genomic DNA with cDNA probes schematically illustrated in Figure 2. The original cDNA clones consisted of a 4 kb EcoRI fragment which encoded the 3'-end of the gene, and a 1 kb EcoRI fragment from the 5'-end (Ullrich et al, 1985). These two cDNA clones have been further subcloned to smaller restriction fragments which have been used to study the 23 different RFLPs observed in various racial groups.

The polymorphisms studied by a number of laboratories in NIDDM subjects and controls have been reviewed (Permutt, 1990). Seven different racial groups were studied. Differences in frequencies of polymorphisms between diabetic and nondiabetic groups were noted in 4 of 11 reports. Two studies found no association of the individual polymorphisms with NIDDM, but did observe positive associations with extended RFLP haplotypes (Xiang et al, 1989; Morgan et al, 1989).

The associations at the insulin receptor locus with NIDDM are examined and compared among studies in Table 2. The most frequently studied RFLPs are two Sst I (Sst-1 and Sst-2) and two Rsa I (Rsa-1 and Rsa-2) RFLPs (Elbein et al, 1986), detected with a single cDNA probe which includes bases 1601-2963 (see Figure 2, Bgl I fragment of HIR 12.1 and comparable probes). The Sst-1 5.8 kb (larger) allele accounted for 13% of alleles in diabetics and 5% in non-diabetics in a mixed Caucasian-Hispanic population (McClain et al, 1988), for 18% of alleles in NIDDM subjects and 6% in non-diabetic British Caucasians, while in contrast the 5.8 kb allele was more common in nondiabetic than in NIDDM subjects in American Blacks (Permutt et al, 1989a). Only the study in American Blacks took into consideration statistical corrections for multiple comparisons, and in that study there were no differences when this correction was employed. Furthermore there were no differences in the frequencies between nondiabetic and diabetic subjects for the 5.8 kb Sst-1 allele in Japanese (Li et al, 1988) or Chinese Americans (Xiang et al, 1989). This particular Sst-1 polymorphism is interesting because it represents a 400 base pair DNA insertion/deletion of an Alu repeat, located between exons 14 and 15 (Elbein, 1989). While it is unlikely that this polymorphic Alu repeat has any functional consequence, it could be in linkage disequilibrium with a mutation, although exons 14,15,16, and 17 have been sequenced from an insertional allele and no mutation found (Elbein, S., personal communication).

Table 2. Associations of Sst-1 and Rsa-1 RFLPs at the Human Insulin Receptor Locus with NIDDM

Racial Group	cDNA Probe	NIDDM	Sst-1 5.3 kb	5.8 kb	6.7 kb	Rsa-1 6.0 kb	3.4 kb	Reference
Japanese	2216-4343	+ (102)	0.84	0.17				Li 1988
		− (100)	0.89	0.11				
Caucasian-Hispanic	2702-3482	+ (51)	0.87	0.13*				McClain 1988
		− (52)	0.95	0.05				
Caucasian	?	+ (56)	0.82	0.18*				Galton 1989
		− (54)	0.94	0.06				
Chinese-American	hINSR101 2180-3018	+ (92)	0.87	0.13	0.13	0.42	0.45	Xiang 1989
		− (74)	0.90	0.10	0.13	0.37	0.50	
American Black	pHIR12.1B1.3 1601-2963	+ (104)	0.73	0.27*	0.27	0.67		Permutt 1989
		− (49)	0.62	0.38	0.30	0.66		
Japanese	pHIR12.1	+ (21)			0.10	0.71	0.86	Takeda 1986
		− (19)			0.21	0.74	0.84	
Mexican American	1601-2963	+ (206)			0.33	0.46	0.22**	Riboudi 1989
		− (185)			0.33	0.50	0.18*	
Pima Indians	pHIR12.1B1.3 1601-2963	Low (24)			0.26	0.52	0.21	Permutt 1989
		High (24)			0.14	0.66	0.20	
Caucasians		+ (131)			0.67	0.33		Rees 1988
		− (72)			0.58	0.42		

* p<0.05; ** N.S., but 3.4/3.4 genotype, p=0.02

(Printed with permission of Balliere Tindall Limited, London.)

The Rsa-1 polymorphism occurs as a three-allele (6.7 kb, 6.0 kb, and 3.4 kb) polymorphism in Orientals (Xiang et al, 1989; Takeda et al, 1986), Mexican-Americans (Raboudi and Frazier, 1989), and Pima Indians (Permutt et al, 1989a), and a two-allele (6.7 kb and 6.0 kb) polymorphism in Caucasians (Ress et al, 1988) and American Blacks (Permutt et al, 1989a) (Table 2). In a population of Mexican-Americans, while the 3.4 kb allelic frequency did not differ, the genotypic frequency was greater in diabetics (Raboudi et al, 1989).

Linkage Analysis at the Insulin Receptor Locus in Families

Very few studies of linkage analysis of NIDDM at the insulin receptor locus have been reported. Linkage was excluded in three large pedigrees with MODY and in a total of four Caucasian families with NIDDM. Cox et al (1989) studied 20 American Black families with at least two sibs with NIDDM. Tight linkage between NIDDM and the insulin receptor locus was ruled out under all genetic models examined. The possibility of a small contribution of this locus as part of the polygenic aspect of the disease has not been excluded by these family studies, however. Furthermore so few studies of families have been conducted that the role of the insulin receptor as a major susceptibility locus in NIDDM has not been adequately assessed.

Direct Analysis by Sequencing Insulin Receptor Genes from Pima Indians

Direct sequencing of the insulin receptor gene which has been amplified by the polymerase chain reaction has been reported(Moller and Flier, 1988). Lymphoblast RNA samples from a Pima and a Pima-Papago Indian with NIDDM were reverse transcribed into cDNA, and seven sets of oligonucleotide primers were used in the polymerase chain reaction to amplify and sequence the cDNAs (Moller et al, 1989). There were no differences between the sequences found and those of normal individuals previously reported. This study contributed information about the coding portion of the receptor gene in at least one allele of each diabetic patient, yet it did not determine whether transcriptional mutations were present. This seems unlikely, however, as the number of insulin receptors on cells from Pima Indians has not been shown to be significantly reduced.

C. Glucose Transporter Genes

Other major candidates which might contribute to the inherited susceptibility to NIDDM are the glucose transporter membrane proteins (Mueckler, 1990; Bell et al, 1990). The best characterized glucose transport protein, isolated from human erythrocyte membranes, was used to obtain a cDNA clone from human hepatoma cells. This cDNA encodes a 492 residue glucose transporter protein which is one member of a family of glucose transporter genes. The gene is about 35 kb in size, encoded by 10 exons, and has been mapped to human chromosome 1p31.1→p35. This gene (Glut-1) encodes the major glucose transporter protein of brain, erythrocytes, and placenta, while the gene is also expressed to a variable extent in all other tissues studied.

A number of polymorphisms have been identified at the Glut-1 locus in various racial groups, and these have been summarized (Permutt, 1990). The most extensive analysis of Glut-1 RFLPs and NIDDM was in American Blacks (Kaku et al, 1990). DNA from 16 individuals was screened with 19 different restriction endonucleases, and four polymorphisms defined. The frequency of the Taq I polymorphism differed between subjects with NIDDM and controls, but not when corrected for multiple observations. Furthermore, this study showed no difference in the frequencies of extended haplotypes between the two groups.

The Xba I polymorphism is a common two allele (6.2 kb and 5.9 kb) RFLP which has been studied by a number of laboratories in 4 racial groups. Two labs reported an association between this RFLP and NIDDM in Caucasians (Li et al, 1990; Baroni et al, 1989) and Japanese (Li et al, 1988). No family studies at the Glut-1 locus have been reported.

Other members of the glucose transporter gene family are also prime candidates for contributing to susceptibility to NIDDM (James et al, 1989; Permutt et al, 1989b). Studies at the liver/islet glucose transporter (Glut-2) and muscle/adipose tissue insulin-responsive glucose transporter (Glut-4) loci have not been reported. We have identified 4 RFLPs at the Glut-2 locus in American Blacks which exhibit a high degree of linkage disequilibrium, and which do not differ in frequency between controls and subjects with NIDDM (Matsutani et al, 1990). At the Glut-4 locus, extensive search revealed only the Kpn I RFLP previously described (Bell et al, 1989). This RFLP was found to occur with the same frequency in nondiabetic and diabetic American Blacks as well (Matsutani et al, 1990).

D. Summary and Analysis of Candidate Gene Studies

Insulin gene mutations contribute to susceptibility to diabetes in rare individuals, yet the results of population association studies and linkage analyses lead one to conclude that defects in insulin genes could at best contribute in only a small fraction of individuals. The same can be said for studies at the other candidate gene loci. The power of these analytical methods, however, needs to be considered.

The ability to detect a difference in the frequency of a genetic marker in NIDDM vs control populations is related to penetrance, the frequency of the marker in the control population, the degree of linkage disequilibrium between the marker and the putative mutation, and the fraction of individuals with NIDDM due to the mutation. Two of these parameters are examined in Table 3, where the frequency of the marker is estimated in 100 nondiabetic and 100 NIDDM subjects. For the purposes of this analysis complete linkage disequilibrium between the marker Al and the mutation, and 100% penetrance are assumed in each case. Only the frequency of the marker A1 in the controls and the fraction of diabetes due to the mutation at the insulin locus are considered. In the first case, the frequency of A1 in the controls is relatively low, and all of the diabetics have the mutation, and the difference in results is highly significant. In the second case, the marker A1 is more common in controls, and here this difference is still highly significant, but less so than if the mutation occurred on the rare allele. Case 3 is somewhat more realistic in that the D mutation occurs on the rare allele, but the mutation only occurs in 20% of diabetics. For this case there is less of a difference than case 2, but still significant. The last case illustrates that if 20% of diabetics have the mutation, and the mutation occurs on the more common allele, there is no significant difference in the frequency of Al in the control and diabetic populations. In summary, this type of analysis clearly illustrates that any gene defect which accounts for only 5-10% of NIDDM would likely not be detected, given the variabilities of incomplete penetrance and the extent of linkage disequilibrium between the markers. The sample sizes studied have been inadequate to detect differences if only 5-10% of diabetics have a defect at a particular candidate locus.

The results of studies in various racial groups at the insulin receptor locus suggest that differences do exist in NIDDM. The insulin receptor is relatively large however (>100kb), and the degree of linkage disequilibrium between a gene marker and a mutation is likely to be less than at the insulin locus (1.4 kb). Common mutations, if they exist in NIDDM, are likely to be heterogeneous (Taylor et al, 1990). Taylor et al (1990) estimated the frequency of insulin receptor mutations in the general population at 1/1000 from the live birth rate of leprechauns. This is likely a minimal estimate, and the number of individuals heterozygous for insulin receptor mutations may be 10-100 times higher. It is apparent, however, that population studies could easily fail to detect associations if 10% of individuals with NIDDM had mutant insulin receptor alleles.

Little data on the role of the Glut-1 (erythrocyte/brain) glucose transporter locus in NIDDM is available, yet current reports suggest that this locus may contribute to genetic susceptibility. The Glut-2 and Glut-4 genes have been studied even less than Glut-1. Like the other candidate genes, however, the role of glucose transporter genes is not likely to be clarified by population studies, or by linkage analysis in unselected families.

Table 3. The likelihood that a genetic marker (Al), in linkage disequilibrium with a diabetes mutation d will be found to occur with a significant difference between a control (n=100) and NIDDM (n=100) group, assuming complete penetrance, and varying 1) the frequency of Al in the control group, and 2) the fraction of NIDDM resulting from the d mutation.

Variables	#1	#2	#3	#4
1. Freq Al in Controls	0.2	0.8	0.2	0.8
2. % of NIDDM with d	1.0	1.0	0.2	0.2
# with Al				
Control	20	80	20	80
NIDDM	100	100	36[*]	84[**]
P	$<10^{-4}$	$<10^{-3}$	$<10^{-2}$	N.S.

[*] $(0.2)(100) + (0.2)(80) = 36$
[**] $(0.8)(100) + (0.2)(20) = 84$

Insulin Promoter - PCR

M   DI   D3   D4

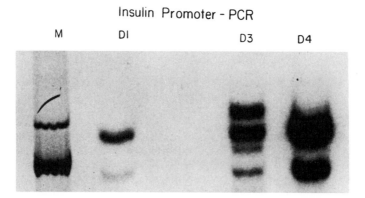

Fig. 3. Amplification of the human insulin gene promoter directly from genomic DNA of NIDDM subjects by the polymerase chain reaction (PCR). DNA (1 ug) was amplified with ^{32}P-labelled oligonucleotide primers (27 bases) spanning the insulin promoter from -360 to +40, and the two strands denatured and electrophoresed on non-denaturing acrylamide gels (Orita et al, 1989). Because this method separates on the basis of size and charge, both strands of the 400 bp promoter can be seen for NIDDM subjects D1 and D4. D3 has this allele and an allele 10 bp larger giving two additional bands. M is the marker, Hinf I digested ^{32}P-labelled plasmid DNA (Bluescript).

IV. USE OF NEW MOLECULAR GENETIC TECHNIQUES

From the results above, it is apparent that alternated methods are required to more precisely define the relationships of these various candidate genes to NIDDM. These include: 1) studying associations with subphenotypes, where the candidate locus ls likely to be involved in a higher % of affected individuals as in first degree relatives of subjects with NIDDM with known insulin resistance, 2) using RFLP haplotypes at each locus, as multiple markers are more likely to discriminate between normal and mutated alleles, and 3) screening populations with simpler and/or more precise analytical methods, such as allele specific oligonucleotide probes for previously described mutations, denaturing gradient gel electrophoresis, single strand conformation polymorphisms (Orita et al, 1989), or sequencing candidate genes directly from genomic DNA. Recently efforts in our lab have been directed to study of the candidate loci with these methods.

A. Direct Genomic Analysis of the Insulin Gene Promoter in NIDDM Subjects

A major hypothesis we have been testing is that the hypoinsulinemia of NIDDM is a consequence of insulin gene transcriptional mutations. Direct analysis of insulin genes in diabetic subjects avoids many of the limitations of restriction fragment length polymorphism (RFLP) analysis. To search for transcription mutants, oligonucleotide primers flanking 375 base pairs of the insulin gene promoter were synthesized and used to amplify the promoter region by the polymerase chain reaction (PCR) directly from genomic DNA of subjects with NIDDM. The advantage of direct genomic analysis over cloning is that the sequences of both alleles can be simultaneously determined. We have confirmed that single base differences in a heterozygous individual can be detected in that we have determined the basis for a previously described Hinc II RFLP at position -56 as GTTGAC to GTTGAG. The promoter region of the insulin gene from 20 American Black subjects with NIDDM was amplified and sequenced, and no difference uncovered. This analysis was satisfying in that it was thorough, but it was labor intensive. We are now developing better screening methods for detecting single base changes such as single strand confirmation polymorphism (Oreta et al, 1989), in which labeled primers are incorporated into PCR amplified products, the strands are denatured and electrophoresed on non-denaturing gels. Mobility is determined by size and base composition, allowing detection of single base changes and small insertion/deletions in short fragments (<300 bp) of DNA. This method allows the ready separation of Hinc II + and - alleles which differ by 1/210 bp. We can then screen the insulin promoter from a large number of diabetic subjects, and sequence only the variants, thus defining more precisely the role of the insulin gene in NIDDM and IDDM. In the initial screening, we detected a 10 bp insertion (Figure 3) in 2/23 NIDDM subjects(2/46 alleles, frequency 0.043). The functional consequence of this insertion is unknown.

B. Correlation of Candidate Gene RFLPs with Subphenotypes

In previous studies we had evaluated the insulin locus in diabetic Pimas with W. Knowler, NIH, Phoenix (Knowler et al, 1984). The analysis at other candidate gene loci in Pimas represents a new collaborative study with C. Bogardus and S. Lillioja, NIH, Phoenix. Nondiabetic Pimas (n=250) who are <35 years of age, and who are expected to develope NIDDM with a very high frequency over the period of observation, are being studied prospectively. This population has been well characterized metabolically, with measures of glucose-stimulated insulin secretion, insulin-mediated whole body glucose disposal, insulin stimulated 2-deoxyglucose uptake in fat cell biopsies, etc.   We plan to correlate these subphenotypes with candidate gene polymorphisms as is being done by others in genetic analysis of the hyperlipidemias.

To date we have prepared DNA and evaluated RFLPs at the Glut-4 locus in 99 Pimas. Analysis of genomic DNA by digestion with the enzyme Kpn I revealed a two allele polymorphism, identical to that described in Caucasians and American Blacks (Matsutani et al, 1990). The frequencies of the smaller +(5.8 kb) and larger -(6.5 kb) alleles in Pimas were very similar to those reported in Caucasians (+ allele 0.37) and in American Blacks (Table 4). There were no differences in the genotypic frequencies for any of the subphenotypic characteristics examined, including % body fat, body mass index, fasting plasma glucose and insulin, plasma glucose and insulin values after glucose stimulation, low dose and high dose glucose utilization values during euglycemic clamps, or basal splanchnic glucose output.

The Glut-4 analysis was done by the traditional Southern blotting and hybridization method. Much of the future analysis will use newer methods described above. For example, we have completed an evaluation of the apoB-100 gene in Pimas, because of the reported association of this gene with NIDDM in Caucasians, and because we can use the hypervariable region approximately 200 bp 3' to the gene on chromosome 2 and PCR (Boerwinkle et al, 1989) to accomplish this analysis far more rapidly than with traditional Southern blotting. A total of 8 different alleles, varying from 28-48 repeat units, were observed in Pima Indians. The frequency distribution of the alleles was very similar to that reported in Austrian Caucasians (Ludwig et al, 1989). The frequency distribution in non-diabetic and diabetic Pimas is seen in Figure 4, but the number of observations of diabetics (n=9) is small. Analysis of apoB-100 genotypes vs subphenotypes is in progress.

Table 4. Allelic Frequencies of Kpn I RFLP at Glut-4

| | Alleles (freq) | | |
	+ (5.8kb)	- (6.5kb)	T
Amer. Blacks	66 (.42)	92 (.58)	158
Pima Indians	78 (.39)	120 (.61)	198

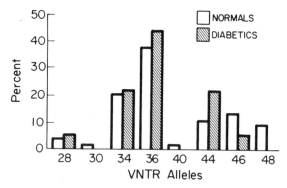

Fig. 4. Amplification of the hypervariable region immediately 3' to the apolipoprotein B gene in unrelated Pima Indians using specific oligonucleotides (Boerwinkle et al, 1989) and the polymerase chain reaction as described (Weber and May, 1989). Amplified DNA containing 28-48 repeats of a basic 15 bp unit were resolved by high resolution polyacrylamide sequencing gel electrophoresis, and compared to standards of known composition provided by E. Ludwig and B. McCarthy (Ludwig et al, 1988). Frequency distribution of hypervariable apoB alleles in 99 Pima Indians, diabetics (n=9) vs nondiabetics.

C. Isolation of Human Glucokinase Genes and Identification of Hypervariable CA Repeats

Recently we isolated a 2.2 kb cDNA clone (hGlk-1) from a human islet library using a rat glucokinase probe (M. Magnusson, Vanderbilt University). The sequence is >90% identical to the rat islet glucokinase in the coding region. Liver and islet isoforms were described in the rat, with different 173 bp exon 1's, and with a 51 bp deletion in exon 4 of the islet cDNA. Interestingly, the 51 bp deletion in exon 4 of the rat islet glucokinase cDNA was not present in hGlk-1, and thus in humans the islet and liver glucokinase are more alike. We used hGlk-1 cDNA to screen a human genomic library (Stratagene), and obtained three overlapping clones comprising a total of 40 kb of genomic DNA. Genomic clones were subcloned into plasmids and intron-exon boundaries are being sequenced. A new method for identifying hypervariable regions in genomic DNA was used to search for polymorphisms at this locus (Weber and May, 1989). The genomic clones were hybridized with poly-dCdA/dGdT to search for CA repeats which have been reported to occur in genomic DNA on the average every 30-60 kb. A single region of CA repeats was mapped to a 1 kb Eco RI fragment in the 3' region of the gene. This region has been subcloned and sequenced, and found to contain 17 CA repeats. Specific oligonucleotide primers surrounding the CA repeats were synthesized, and the region evaluated in genomic DNA of individuals by PCR methods. At least two polymorphic alleles have been identified in three racial groups. This promises to be an important finding for mapping the chromosomal location of the gene, and for evaluating the role of defects at this locus in NIDDM.

SUMMARY

Polymorphisms, both restriction fragment length polymorphisms (RFLPs) and variable number of tandem repeats (VNTRS) occur commonly in human genomic DNA, and these polymorphisms provide useful markers for genetic analysis. Polymorphisms have been used to evaluate specific candidate loci for NIDDM, e.g. the insulin, insulin receptor, and glucose transporter genes. For these analyses, population and family studies (limited in number) have suggested that none of these loci are major contributors to the genetic susceptibility to NIDDM. Analysis is complicated if NIDDM is polygenic, however. In no case could a contribution of 10% or less of these loci be confidently excluded, because of variable penetrance, different degrees of linkage disequilibrium between RFLPs and putative mutations, the frequencies of the RFLPs in non-diabetic populations, and inadequate sample size. These results lead one to conclude: either 1) the correct candidate gene(s) has not been found, or 2) sample sizes need to be increased by at least an order of magnitude, or 3) newer methods of analysis must be adopted (e.g. use of extended haplotypes and associations with subphenotypes, or more direct analysis of genomic DNA.

ACKNOWLEDGEMENTS

We would like to acknowledge the many contributions of our colleagues Steve Elbein, Mike Province, Janet McGill, Kohei Kaku, Mike Mueckler, Clifton Bogardus, and Stephen Lillioja. The laboratory assistance of Ms. Cris Welling and Rachel Janssen, and manuscript preparation by Ms. Jeannie Wokurka, are also gratefully acknowledged. Part of the work presented was supported by an NIH grant (R37DK16746), a Juvenile Diabetes Foundation Fellowship (Laszlo Koranyi), and an American Diabetes Mentor-Based Fellowship (Akira Matsutani).

REFERENCES

Accili, D., Frapier, C., Mosthaf, L., et al, 1989, A mutation in the insulin receptor gene that impairs transport of the receptor to the plasma membrane and causes insulin resistant diabetes, EMBO Journal, 8:2509.

Antonarakis, S.E., Oettgen, P., Chakravarti, A., et al, 1988, DNA polymorphism haplotypes of the human apolipoprotein AP0A1-AP0C3-AP0A4 gene cluster, Hum Genet, 80:265.

Aoyama, N., Nakamura, T., Doi, K., et al, 1986, Low frequency of the 5'-flanking insertion of human insulin gene in Japanese non-insulin dependent diabetic subjects, Diabetes Care, 9:365.

Awata, T., Shibasaki, Y., Hirai, H., et al, 1985, Restriction fragment length polymorphism of the insulin gene region in Japanese diabetic and non-diabetic subjects, Diabetologia, 28:911.

Barker, P.E., 1990, Gene-mapping and cystic-fibrosis, Amer J Med Sci, 299:69.

Baroni, M.G., Pozzilli, P., Oelbaum, R.S., Li, S.R., and Galton, D.J., 1989, A polymorphic site in the glucose transporter gene associates with the type 2 (non-insulin-dependent) diabetes mellitus in the Italian population, Diabetologia, 32:464.

Bell, G.I., Karam, J.H., and Rutter, W.J., 1981, Polymorphic DNA region adjacent to the 5' end of the human insulin gene, Proc Natl Acad Sci USA, 78:5759.

Bell, G.I., Selby, M.J., and Rutter, W.J., 1982, The highly polymorphic region of the human insulin gene is composed of simple tandemly repeating sequences, Nature, 295:31.

Bell, G.I., Kayano, T., Buse, J.B., et al, 1990, Molecular biology of mammalian glucose transporters, Diabetes Care, 13:198.

Bell, G.I., Murray, J.C., Nakamura, Y., et al, 1989, Polymorphic human insulin-responsive glucose transporter gene expressed in skeletal muscle and other insulin-responsive tissues is on chromosome 17p13, Diabetes, 38:1072.

Boerwinkle, E., Xiong, W., Fourest, E., and Chan, L., 1989, Rapid typing of tandemly repeated hyper-variable loci by the polymerase chain reaction: Application to the apolipoprotein B 3' hypervariable region, Proc Natl Acad Sci USA, 86:212.

Chakravarti, A., Elbein, S.C., and Permutt, M.A., 1986, Evidence for increased recombination near the human insulin gene: Implication for disease association studies, Proc Natl Acad Sci USA, 83:1045.

Coleman, D.L., and Hummel, K.P., 1975, Influence of genetic background on the expression of mutations at the diabetes locus in the mouse. II. Studies on background modifiers, Israel J Med Sci, 11:708.

Cox, N.J., and Spielman, R.S., 1989, The insulin gene and susceptibility to IDDM, Genetic Epidemiology, 6:65.

Cox, N.J., Baker, L., and Spielman, R.S., 1988, Insulin-gene sharing in sib pairs with insulin dependent diabetes mellitus: No evidence for linkage, Amer J Hum Genet, 42:167.

Cox, N.J., Bell, G.I., and Xiang, K., 1988, Linkage disequilibrium in the human insulin/insulin-like growth factor II region of human chromosome II, Amer J Hum Genet, 43:495.

Cox, N.J., Epstein, P.A., and Spielman, R.S., 1989, Linkage studies on IDDM and the insulin and insulin-receptor genes, Diabetes, 38:653.

DeFronzo, R.A., 1988, Lilly Lecture 1987. The triumvirate: β-cell,muscle, liver: A collusion responsible for NIDDM, Diabetes, 37:667.

Ebina, Y., Ellis, L., Jarnagin, K., et al, 1985, The human insulin receptor cDNA: The structural basis for hormone-activated transmembrane signalling, Cell 40:747.

Elbein, S.C., 1989, Molecular and clinical characterization of a insertional polymorphism of the insulin receptor gene, Diabetes, 38:737.

Elbein, S.C., Ward, W.K., Beard, J.C., and Permutt, M.A., 1988, Familial NIDDM: Molecular-genetic analysis and assessment of insulin action and pancreatic β-cell function, Diabetes, 37:377.

Elbein, S.C., Corsetti, L., Goldgar, D., Skolnick, M., and Permutt, M.A., 1988, Insulin gene linkage in familial NIDDM: Lack of linkage in Utah Mormon pedigrees, Diabetes, 37:569.

Elbein, S., Rotwein, P., Permutt, M.A., et al, 1985, Lack of association of the polymorphic locus in the 5' flanking region of the human insulin gene and diabetes in American Blacks, Diabetes, 34:433.

Eriksson, J., Fransilakallunki, A., Ekstrand, A., et al, 1989, Early metabolic defects in persons at increased risk for non-insulin-dependent diabetes-mellitus, New England Journal of Medicine 321:337.

Fajans, S.S., 1990, Scope and heterogeneous nature of MODY, Diabetes Care, 13:49.

Frazier, M.L., Ferrell, R.E., Arem, R., Chamakhi, S., and Field, J.B., 1986, Restriction fragment length polymorphism of the human insulin gene region among type II diabetic Mexican-Americans and Tunisians, Res Comm Chem Pathol Pharmacol, 52:371.

Galton, D.J., 1989, DNA polymorphisms of the human insulin and receptor genes in type 2 diabetes mellitus, Diab/Metab Rev, 5:443.

Gruppuso, P.A., Gorden, P., Kahn, C.R., Cornblath, M., Zeller, W.P., and Schwartz, R., 1984, Familial hyperproinsulinemia due to a proposed defect in conversion of proinsulin to insulin, New Engl J Med, 311:629.

Haneda, M., Kobayashi, M., Maegawa, H., and Shigeta, Y., 1986, Low frequency of the large insertion in the human insulin gene in Japanese, Diabetes, 35:115.

Haneda, M., Polonsky, K.S., Bergenstal, R.M., et al, 1984, Familial hyperinsulinemia due to a structurally abnormal insulin: Definition of an emerging new clinical syndrome, New Engl J Med, 310:1288.

Hitman, G.A., Karir, P.K., Mohan, V., et al, 1985, A genetic analysis of type 2 diabetes mellitus in Punjabi Sikhs and British Caucasoid patients, Diab Med, 4:526.

James, D.E., Strube, M., and Mueckler, M., 1989, Molecular cloning and characterization of an insulin regulatable glucose transporter, Nature, 338:83.

Kaku, K., Province, M., and Permutt, M.A., 1989, Genetic analysis of obesity-induced diabetes associated with a limited capacity to synthesize insulin in C57BL/KS mice: Evidence for polygenic control, Diabetologia, 32:636.

Kaku, K., Matsutani, A., Mueckler, M., and Permutt, M.A., 1990, Polymorphisms of HepG2/erythrocyte glucose-transporter gene: Linkage relationships and implications for genetic analysis of NIDDM, Diabetes, 39:49.

Kanazawa, Y., Awata, T., Shibasaki, Y., and Takaku, F., 1988, Polymorphism of insulin gene area in Japanese: Relationship with insulin-dependent and non-insulin dependent diabetes, Biomed Biochim Acta, 47:317.

Knowler, W.C., Pettitt, D.J., Savage, P.J., and Bennett, P.H., 1981, Diabetes incidence in Pima Indians: Contributions of obesity and parental diabetes, Am J Epidemiol, 113:144.

Knowler, W.C., Pettitt, D.J., Vasquez, B., et al, 1984, Polymorphism in the 5' flanking region of the human insulin gene: Relationships with non-insulin-dependent diabetes mellitus, glucose and insulin concentrations and diabetes treatment in the Pima Indians, J Clin Invest, 74:2129.

Lehrman, M.A., Goldstein, J.L., Russell, D.W., and Brown, M.S., 1987, Duplication of seven exons in LDL receptor gene caused by Alu-Alu recombination in a subject with familial hypercholesterolaemia, Cell, 48:827.

Leiter, E.H., Le, P.H., and Coleman, D.L., 1987, Susceptibility to db gene and streptozotocin-induced diabetes in C57BL mice: Control by gender-associated, MHC-unlinked traits, Immunogen, 26:6.

Li, S.-R., Oelbaum, R.S., and Galton, D.J., 1988, DNA polymorphisms of the insulin receptor gene in Japanese subjects with non-insulin-dependent diabetes mellitus, Hum Hered, 38:273.

Li, S.-R., Oelbaum, R.S., Bouloux, P.M.G., Stocks, J., Baroni, M.G., and Galton, D.J., 1990, Restriction site polymorphisms at the human HepG2 glucose transporter gene locus in Caucasian and West Indian subjects with non-insulin-dependent diabetes mellitus, Hum Hered, 40:38.

Ludwig, E.H., Friedl, W., and McCarthy, B.J., 1988, High-resolution analysis of a hyper-variable region in the human apolipoprotein B gene, Am J Hum Genet, 45:458.

Lusis, A.J., 1988, Genetic factors affecting blood lipoproteins: The candidate gene approach, J Lipid Res, 29:397.

Matsutani, A., Koranyi, L., Cox, N., and Permutt, M.A., 1990, Polymorphisms of the liver/islet (Glut-2) and muscle/adipose (Glut-4) glucose transporter genes: Use in genetic evaluation of genetic susceptibility to NIDDM in American Blacks, Diabetes, in press.

McClain, D.A., Henry, R.R., Ullrich, A., and Olefsky, J.M., 1988, Restriction-fragment-length polymorphism in insulin-receptor gene and insulin resistance in NIDDM, Diabetes 37:1071.

Moller, D.E., and Flier, J.S., 1988, Detection of an alteration in the insulin-receptor gene in a patient with insulin resistance, acanthosis nigricans, and the polycystic ovary syndrome (type A insulin resistance), New Engl J Med, 319:1526.

Moller, D.E., Yokota, A., and Flier, J.S., 1989, Normal insulin-receptor cDNA sequence in Pima Indians with NIDDM, Diabetes, 38:1496.

Morgan, R., Bishop, A., Sarfarazi, M., et al, 1989, Specific insulin receptor gene haplotypes which associate with type 2 (non-insulin-dependent) diabetes, Diabetologia, 32:519.

Mueckler, M., 1990, Family of glucose-transporter genes: Implications for glucose homeostasis and diabetes, Diabetes, 39:6.

Nagulesparan, M., Savage, P.J., Knowler, W.C., Johnson, G.C., and Bennett, P.H., 1982, Increased in vivo insulin resistance in nondiabetic Pima Indians compared with Caucasians, Diabetes, 31:592.

O'Malley, K.I., and Rotwein, P., 1988, Human tyrosine hydroxylase and insulin genes are contiguous on chromosome 11, Nucl Acids Res, 16:4437.

O'Rahilly, S., Trembath, R.C., Patel, P., et al, 1988, Linkage analysis of the human insulin-receptor gene in type-2 (non-insulin-dependent) diabetic families and a family with maturity onset diabetes of the young, Diabetologia, 31:792.

Orita, M., Suzuki, Y., Sekiya, T., and Hayashi, K., 1989, Rapid and sensitive detection of point mutations and DNA polymorphisms using the polymerase chain reaction, Genomics, 5:874.

Owerbach, D., Bell, G.I., Rutter, W.J., and Shows, T.B., 1980, The insulin gene is located on chromosome 11 in humans, Nature, 286:82.

Permutt, M.A., 1990, The use of DNA polymorphisms for genetic analysis of non-insulin dependent diabetes, in: "The Genetics of Diabetes," B.D. Tait, L.C. Harrison, eds., Balliere Tindall Limited, London, in press.

Permutt, M.A., and Elbein, S.C., 1990, Insulin gene in diabetes: Analysis through RFLP, Diabetes Care, 13:364.

Permutt, M.A., McGill, J., Elbein, S.C., Province, M., and Bogardus, C., 1989, Insulin receptor gene polymorphisms (RFLPs) in American Blacks and Pima Indians: An assessment of the use of RFLPs in evaluating a candidate locus for NIDDM, in "Nordisk Insulin Symposium No. 3. Genes and Gene Products in the Development of Diabetes Mellitus. Basic and Clinical Aspects," J. Nerup, T. Mandrup-Poulsen, and B. Hokfelt, eds., Elsevier Science Publishers B.V., Amsterdam.

Permutt, M.A., Koranyi, L., Keller, K., et al, 1989, Cloning and functional expression of a human pancreatic islet glucose transporter cDNA, Proc Natl Acad Sci USA, 86:8688.

Raboudi, S.H., and Frazier, M.L., 1989, Restriction fragment length polymorphism of the human insulin receptor gene among Mexican Americans, Res Comm Chem Pathol Pharmacol, 65:319.

Raboudi, S.H., Mitchell, B.D., Stern, M.P., et al, 1989, Type II diabetes mellitus and polymorphism of insulin-receptor gene in Mexican Americans, Diabetes, 38:975.

Rees, A., Morgan, R., Bishop, A., et al, 1988, Insulin receptor gene variants in type 2 (non-insulin-dependent) diabetes mellitus, Diabetologia, 37:534A.

Rotter, J.I., Vadheim, C.M., and Rimoin, D.L., 1990, Genetics of diabetes mellitus, in "Diabetes Mellitus, Theory and Practice, Fourth Edition," H. Rifkin, and D. Porte, Jr., eds., Elsevier Science Publishing Co., New York.

Seino, S., Seino, M., and Bell, G.I., 1990, Human insulin-receptor gene, Diabetes 39:129.

Serjeantson, S.W., Owerbach, D., Zimmet, P., Nerup, J., and Thoma, K., 1983, Genetics of diabetes in Aurau: Effects of foreign admixture, HLA antigens and the insulin-gene-linked polymorphism, Diabetologia, 25:13.

Spielman, R.S., Baur, M.P., and Clerget-Darpoux, F., 1989, Genetic analysis of IDDM: Summary of GAW5 IDDM results, Genetic Epidemiology, 6:43.

Taira, M., Taira, M., Hashimoto, N., et al, 1989, Human diabetes associated with a deletion of the tyrosine kinase domain of the insulin receptor, Science, 245:63.

Takeda, J., Seino, Y., Fukumoto, H., et al, 1986, The polymorphism linked to the human insulin gene: Its lack of association with either IDDM or NIDDM in Japanese, Acta Endocrinol (Copen), 113:268.

Taylor, S.I., Kadowaki, T., Kadowaki, T., et al, 1990, Mutations in insulin-receptor gene in insulin-resistant patients, Diabetes Care, 13:257.

Trucco, M., and Dormon, J.S., 1989, Immunogenetics of insulin dependent diabetes mellitus in humans, Crit Rev Immunol, 9:201.

Ullrich, A., Dull, T.J., and Gray, A., 1980, Genetic variation in the human insulin gene, Science, 200: 612.

Ullrich, A., Bell, G., Chen, E.Y., et al, 1985, Human insulin receptor and its relationships to the tyrosine kinase family of oncogenes, Nature, 313:756.

Weber, J.L., and May, P.E., 1989, Abundant class of human DNA polymorphisms which can be typed using the polymerase chain reaction, Am J Hum Genet, 44:388.

Xiang, K., and Bell, G.I., 1988, Apa I and Sst I RFLPs at the insulin-like growth factor II (IGF2) locus on chromosome ll, Nucl Acids Res, 16:3599.

Xiang, K., Karam, J.H., and Bell, G.I., 1987, BamHI RFLP at the insulin-like growth factor II (IGF2) locus on chromosome ll, Nucl Acids Res, 15:7655.

Xiang, K.-S., Cox, N.J., Sanz, N., et al, 1989, Insulin-receptor and apolipoprotein genes contribute to development of NIDDM in Chinese Americans, Diabetes, 37:17.

Zimmet, P., 1982, Type 2 (non-insulin-dependent) diabetes - An epidemiological overview, Diabetologia, 22:399.

HDL AND REVERSE TRANSPORT OF CHOLESTEROL:
INSIGHTS FROM MUTANTS

Gerd Assmann,[1,2] Arnold von Eckardstein,[1] and Harald Funke[1]

[1]Institut für Klinische Chemie und Laboratoriumsmedizin
Zentral-Laboratorium
Westfälische Wilhelms-Universität Münster
Albert-Schweitzer-Str. 33
D-4400 Münster, F.R.G.

[2]Institut für Arterioskleroseforschung an der Universität Münster
Domagkstr. 3
D-4400 Münster, F.R.G.

1 Introduction

Several epidemiological and clinical studies revealed an inverse correlation between low plasma concentrations of high density lipoprotein (HDL) cholesterol as well as its major protein component apolipoprotein A-I (apo A-I) and the risk of myocardial infarction (reviewed in 1). Family and twin studies suggested partial heredity of low HDL-cholesterol levels and have put the influence of genes at 35 to 50% (2,3). Frequently, familial HDL-cholesterol-deficiency was paralleled with a family history of premature coronary heart disease (CHD) (4,5).

The pathophysiological relationship between decreased serum concentrations of HDL-cholesterol and the development of coronary heart disease, however, has still remained obscure. The reverse cholesterol transport model (6) is most widely used to explain the role of HDL in cellular lipid metabolism and in atherogenesis. In this model, HDL mediates the flux of excess cholesterol from peripheral cells to the liver.

Genetic defects that interfere with the regular structure of HDL apolipoproteins, of the quaternary structure of HDL, or with processes important of the generation or removal of HDL, are of particular importance to gain insight into the function of these lipoproteins (7,8)

Correspondence should be addressed to: Prof. Dr. med. G. Assmann, Institut für Klinische Chemie und Laboratoriumsmedizin Zentral-Laboratorium, Westfälische Wilhelms-Universität Münster, Albert-Schweitzer-Str. 33, D-4400 Münster, F.R.G.

2 The reverse cholesterol transport model

Different subpopulations of HDL can be differentiated by ultracentrifugation, by their mobility in agarose gel electrophoresis, by isotachophoresis or by their apolipoprotein composition. They likely take up excess cellular cholesterol by at least two different mechanisms (reviewed in 3,7-9): (i) Unesterified cholesterol from cell membranes is removed by nascent, solely apo A-I containing HDL particles with preß mobility on agarose gel electrophoresis and further esterified by the plasma enzyme lecithin:cholesterol-acyltransferase (LCAT). Nascent HDL, discoidal in shape and lipid-poor particles, are generated by direct secretion from the liver as well as through lipolysis of chylomicrons and VLDL, and have been termed preß-HDL, Lp A-I w/o A-II, LpA-I$_{preß}$, or slow migrating HDL. Cholesteryl esters (CE) formed by LCAT are packed into the hydrophobic core of these lipoproteins, thus enlarging the particles and making them spherical. Parallel with this process, LpA-I$_{preß}$ acquire apo D and the cholesterol ester transfer protein (CETP). (ii) Certain subpopulations of HDL also interact with specific cell surface recognition sites ("HDL-receptors") either by signal transduction or by retroendocytosis and thereby promote cholesterol efflux from these cells. In both pathways ongoing cholesterol esterification via LCAT activity results in the formation of spherical HDL$_3$ that serve as acceptors of lipoprotein surface remnants generated by the enzymatic hydrolysis of triglycerides contained in chylomicrons and VLDL. This process and continuing LCAT action converts HDL$_3$ to HDL$_2$. From various HDL subpopulations CE can be removed by at least four different mechanisms and ultimately taken up by the liver where cholesterol is utilized for bile acid synthesis: (i) A subpopulation of HDL acquires apo E and can be recognized by hepatic apo E receptors. (ii) HDL without apo E can also be endocytosed by hepatocytes. (iii) Hepatic triglyceride lipase (HTGL) mediates the uptake of cholesterol esters into liver cells. (iv) CETP catalyzes the net transfer of HDL-cholesterol esters to LDL and VLDL which are then taken up via hepatic apo B,E receptors (7).

Genetic variations causing structural defects of HDL apolipoproteins (A-I, A-II, A-IV, and C-III) or of lipid transfer enzymes (CETP and LCAT) importantly affect the plasma concentration of HDL cholesterol as well an the effectiveness of reverse cholesterol transport.

3 Mutations in apolipoprotein genes

3.1 Apo A-I deficiency and apo A-I mutants

HDL cholesterol and apo A-I were absent or severely reduced in patients from six families because of mutations in the apo A-I gene that either prevented synthesis or led to the secretion of structurally grossly changed proteins (10-17, table 1). The clinical presentation of these patients was variable and included premature atherosclerosis, xanthomatosis, and corneal opacities. Premature atherosclerosis, however, was only demonstrated in two cases, in which an inversion or a deletion prevented the synthesis of

Table 1. Molecular Defects in the to Apo A-I Gene

molecular defect	clinical and functional consequences

<u>Defects leading to HDL deficiency</u>

molecular defect	clinical and functional consequences
deletion of APOLP1	apo A-I/C-III/A-IV deficiency
	xanthomatosis, CHD
	partial LCAT deficiency
inversion within APOLP1	apo A-I/C-III-deficiency
	xanthomatosis, CHD
single nucleotide	apo A-I deficiency, secretion of a frameshift protein
deletion in codon 202	corneal opacities
	partial LCAT deficiency
single nucleotide	apo A-I deficiency
insertion in codon 3-5	xanthomatosis, CHD
deletion of residues 146-160	apo A-I deficiency
	partial LCAT deficiency
$Gln_{32}->STOP$	apo A-I deficiency
	PARTIAL lcat DEFICIENCY

<u>Apo A-I Mutants</u>

molecular defect	clinical and functional consequences
$Pro_3->Arg$	impaired proapo A-I conversion
$Pro_3->His$	impaired proapo A-I conversion
$Pro_4->Arg$	
$Arg_{10}->Leu$	
$Asp_{13}->Tyr$	
$Gly_{26}->Arg$	familial polyneuropathy, low HDL
$Asp_{89}->Glu$	
$Asp_{103}->Asn$	
$Lys_{107}->0$	reduced LCAT cofactor activity
	impaired lipid binding
$Lys_{107}->Met$	
$Glu_{110}->Lys$	
$Glu_{136}->Lys$	
$Glu_{139}->Gly$	
$Pro_{143}->Arg$	reduced LCAT cofactor activity
$Glu_{147}->Val$	
$Ala_{158}->Glu$	
$Pro_{165}->Arg$	reduced LCAT cofactor activity
	low HDL cholesterol
$Glu_{169}->Gln$	
$Arg_{173}->Cys$	low HDL cholesterol
	partial LCAT deficiency
	abnormal HDL particles
	altered lipid binding properties
$Arg_{177}->His$	
$Glu_{198}->Lys$	
$Asp_{213}->Gly$	

For origin see references 3, 14-17

both apo A-I and apo C-III, or of apolipoproteins A-I, C-III and A-IV, respectively (12,13) By contrast, homozygosity for four other mutations in the apo A-I gene severely reduced HDL cholesterol plasma concentrations but did not appear to put affected individuals at increased coronary risk (14-17). Also family histories were not in support of any increased prevalence of myocardial infarction within these kindreds.

In several cases the impact of absent HDL on plasma LCAT activity was analysed (10,14,15,17,18): In all homozygous individuals both LCAT activity and LCAT mass were decreased. Partial LCAT deficiency in the affected individuals, therefore, is likely secondary to the absence of small HDL and, by implication, these small HDL particles appear to play a role in the metabolism of the enzym LCAT (10)

Screening studies in 32 000 German newborns and analyses of patients with low HDL cholesterol by isoelectric focusing (IEF) helped to identify a large number of structural apo A-I variants (reviewed in 3,19. table 1). Most of these apo A-I variants did not affect plasma lipid and lipoprotein concentrations, but at least three mutants have been associated with decreased HDL-cholesterol plasma consentrations in the heterozygous carriers: apo A-I_{Milano}(Arg$_{173}$-Cys), apo A-I(Pro$_{165}$-Arg) and apo A-I_{Iowa}(Gly$_{26}$-Arg). However, in none of these three probands or their families premature CAD could be demonstrated. The apo A-I (Gly$_{26}$-Arg) carrier was affected by amyloidotic polyneuropathy (20,21). In the Apo A-I_{Milano} carrier and several family members major anomalies in the structure and metabolism of HDL apolipoproteins (Apo A-I homodimers, apo A-I-A-II heterodimers, impaired lipid binding, presence of atypical HDL subpopulations) and partial LCAT deficiency were demonstrated. In heterozygous apo A-I (Pro$_{165}$-Arg) carriers the relative concentration of the mutant apo A-I was found decreased to 30 percent (instead of the expected 50 percent) and apo A-I deficiency could be attributed to the decreased concentration of the variant apo A-I (24). A similar phenomenon has been observed in apo A-I (Pro$_{143}$-Arg) carrier (25). After isolation and reconstitution of apo A-I (Pro$_{143}$-Arg) and apo A-I (Pro$_{165}$-Arg) with various phospholipids (dimyristoylphosphatidylcholine, dipalmitoylphosphatidylcholine, palmitoyloleoylphosphatidylcholine) these variants were defective in activating LCAT (25,26). Using the Chou-Fasman algorithm both the Pro$_{143}$-Arg and the Pro$_{165}$-Arg substitution caused the elimination of beta-turns between two neighbouring alpha-helices and apparently interfered with important requirements for regular LCAT activation and lipid binding. Lipid binding LCAT activation, and tryptophan fluorescence properties were also altered in apo A-I(Lys$_{107}$-O) carriers, but in extensive family studies (nine families) all heterozygous family members were not affected by dyslipoproteinemia (19,26-29).

Apo A-I is secreted as proapo A-I and maturated by extracellular cleavage of an aminoterminal hexapeptide by a hitherto unknown propeptidase. Studies in three apo A-I variants in which proline residues 3 or 4 were substituted by other amino acids revealed structural requirements for the regular processing of apo A-I. In heterozygous carriers of apo A-I(Pro$_3$-Arg) and apo A-I(Pro$_3$-His) the concentration of the variant proapo A-I was found relatively increased compared to normal proapo A-I. This observation was not made in apo A-I(Pro$_4$-Arg) carriers. These data suggested that only proline residue 3 which is *inter species* highly conserved, is important for the regular conversion of proapo A-I (24).

3.2 Mutations in the genes for apo A-II, apo A-IV, and apo C-III

Besides apo A-I, other proteins (apo A-II, apo A-IV, apo C's, apo D, apo E and apo J) are likely important structural components of HDL. To date, only few mutants of these apolipoproteins have been identified that may affect HDL metabolism. Interestingly, a proband with apo A-II deficiency exhibited normal plasma concentrations of lipoproteins including HDL cholesterol (30).

Upon isoelectric focusing, apo A-IV exhibits a genetic polymorphism. In the Caucasian population the frequency of the most frequent allele - apo A-IV-1 - is approximately 92%, the frequency of the second frequent allele - apo A-IV-2 - 7-8%. Other isoforms designated apo-A-IV-3, apo A-IV-4 and apo A-IV-0 are very rare. The molecular basis of the charge difference between the two more frequent alleles - apo A-IV-1 and apo A-IV-2 - is a G to T substitution in codon 360 leading to the replacment of glutamine by histidine (31). The impact of the apo A-IV polymorphism is controversely discussed: In Tyroleans and Icelanders, apo A-IV-1/2 heterozygotes were found to exhibit higher HDL-cholesterol and lower triglyceride plasma concentrations than apo A-IV-1/1 homozygotes (32,33). Other investigators did not find these differences. Our own studies in students, investing at the DNA level these codon differences, failed to establish differences in HDL-cholesterol levels, but showed 10% higher LDL-cholesterol levels in apo A-IV-1/2 heterozygotes compared to apo A-IV-1/1 homozygotes (von Eckardstein, Funke and Assmann, manuscript in preparation). *In vitro*, apo A-IV-2 is more efficient in LCAT activation and phospholipid binding than apo A-IV-1 (34). Upon sequence analysis of the apo A-IV gene in several individuals several other polymorphic sites were detected , some of which lead to electrophoretically undetectable amino acid substitutions. The mutation most frequent in Caucasians leads to a Ser-Thr replacement in position 347. The allele frequencies were 84% (Thr$_{347}$) and 16% (Ser$_{347}$), respectively (35). A study in 300 students revealed that apo A-IV(Ser$_{347}$) is associated with lower apo B and LDL-cholesterol and higher HDL-cholesterol concentrations (36).

An apo C-III variant - apo C-III(Lys$_{58}$-Glu) - was found in a woman and her mother with HDL-cholesterol and apo A-I plasma concentrations exceeding the 95th percentiles of sex matched controls. The unaffected father and sister exhibited normal values. Hyperalphalipoproteinemia in this family was accompanied by decreased apo C-III plasma concentrations in the two variant carriers. The low serum concentration of apo C-III appeared to be a consequence of diminished concentrations of the variant apo C-III isoproteins in both VLDL and HDL. Both the decreased apo C-III concentration and the structural defect in apo C-III possibly account for disinhibited lipolysis and, since surface remnants of triglyceride rich particles are precursors of HDL, for the observed hyperalphalipoproteinemia (37).

4 Lipid tranfer enzymes

4.1 Familial LCAT deficiency and fish eye disease

Familial LCAT deficiency and fish eye disease (FED) are two clinical conditions that are characterized by severely reduced plasma concentrations of HDL cholesterol and by impaired esterification of serum cholesterol. However, they differ by clinical symptoms and by qualitatively and quantitatively different disturbances of cholesterol esterification: Clinically, patients with FED suffer from massive corneal opacifications that provided the unusual name of this disease. In patients with familial LCAT deficiency nephropathy of varying severity (proteinuria, hematuria, renal insufficiency), anemia, and atheroslerosis are major clinical complications. In FED, LCAT activity is undetectable both in HDL or apo A-I containing proteoliposomes but normal in VLDL and LDL, resulting in near normal *in vivo* ratios of unesterified / esterified cholesterol. By contrast, in familial LCAT deficiency esterification of cholesterol is disturbed in all lipoprotein classes and the ratio unesterified / esterified cholesterol is severely increased (reviewed in 10,38). To date, structural analysis of the LCAT genes in 6 families with familial LCAT deficiency (reviewed in 10,39,40) and in two families with FED (41) revealed a variety of mutations in the LCAT gene for both conditions (table 2). None of the reported mutations affected the putative catalytic center of LCAT. Thus, on the molecular level, the difference beween FED and LCAT deficiency appears to lie in the specifically disturbed potency of the LCAT enzyme to bind to or to be active in different lipoproteins. The differently disturbed LCAT activity leads to different changes in the composition and size of lipoproteins: in particular, LDL are grossly changed in familial LCAT deficency but normal in FED. Likely, the failure to esterify cholesterol in triglyceride-rich lipoproteins (VLDL, IDL, remnant particles) has major clinical

Table 2. Molecular defects in the LCAT gene

Origin	Structural changes
familial LCAT deficiency:	
Italy	homozygosity for $Thr_{321}Ile$
Canada	comp. heterozygosity for $Arg_{135}Trp$ and frameshift at residue 375
Italy	comp. heterozygosity for $Arg_{147}Trp$ and a second unknown defect ?
Denmark	homozygosity for $Ala_{93}Thr$ and $Arg_{158}Cys$
France	homozygosity for $Lys_{209}Pro$
fish-eye disease:	
Germany and Netherlands:	homozygosity for $Thr_{123}Ile$

For reference see 9

implications in LCAT deficiency (e.g. nephropathy). Although in both FED and LCAT deficiency HDL is extremely low in concentration and highly abnormal in subpopulation distribution, risk for atherosclerosis in homozygous individuals may only be moderately increased and, at least in LCAT deficiency, partly due to renal disease. Heterozygous family members in both disorders exhibit HDL cholesterol concentrations often lower than 5th percentile of sex matched controls, but family histories do not account for any increased prevalence of myocardial infarction in these kindreds.

4.2 CETP deficiency

CETP deficiency has been elucidated as the metabolic origin of hyperalphalipoproteinemia in several Japanese families (reviewed in 42). In one family with CETP deficiency, the molecular defect has been established as a GA transversion in the 5'-splice donor site of exon 14 of the CETP gene (43). Due to the absence of CETP activity, cholesterol esters formed by LCAT accumulate in HDL. These transform to apo E-rich, HDL_c-like particles and appear to be recognized by LDL receptors. Patients with CETP deficiency have no signs of atherosclerotic diseases. Moreover, this condition was suggested to be a cause for longevity (44).

5 Open questions

Familial HDL deficiency includes a wide variety of disorders of different etiology and different phenotypic expression. To date, the molecular basis of several disorders including LCAT deficiency, FED and apo A-I deficiency have been elucidated. In other disorders of unknown origin, e.g. in Tangier disease, alterations in the plasma lipid metabolism possibly are secondary to disturbances in the intracellular lipid metabolism. Although familial HDL deficiency is very rare, the biochemical and molecular characterization of various defects will help to understand both the origin of the different phenotypic expression of HDL deficiency syndromes and the importance of the different hypothetic functions of HDL in lipid metabolism: reverse cholesterol transport, intercellular cholesterol trafficking, and cellular signalling.

The most important and as yet unanswered question relates to the risk of premature CHD in familial HDL deficiency syndromes. Mutations in the apo A-I gene were associated with increased CHD risk in only few patients. Paradoxically, the absence of normal apo A-I from the plasma compartment in certain individuals did not appear to be associated with vascular anomalies. Likewise, in familial LCAT deficiency and FED severe reduction of HDL cholesterol is not associated with a major risk for premature CHD. In view of nearly absent HDL cholesterol in these conditions, the discrepancy with the epidemiological findings addresses the question of the proposed causative role of HDL in preventing premature atherosclerosis. Possibly the epidemiologic data need other interpretations than provided by the reverse cholesterol transport model, and low HDL-cholesterol only indicates other atherogenic disturbances, e.g. in the metabolism of triglyceride rich particles.

References

1.   Gordon, D and Rifkind, BM. *N. Engl. J. Med.* 1989. **321**:1311-1315.

2.   Hunt, SC, Hasstedt, SJ, Kuida, H, Stults, BM, Hopkins, PN and Wiliams, RR. *Am. J. Epidemiology* 1989. **129**:625-638.

3.   Assmann, G, Schmitz, G, Funke, H and von Eckardstein, A Current opinion in lipidology. 1990. 1:110-115.

4.   DeBacker, G, Hulstaerdt, F, DeMunck, Rosseneu, M, Van Parijs, L and Dramaix, M. *Am Heart J.* 1986. **112**:478-484.

5.   Pometta, D, Micheli, H, Suenram, A, Jornot, C. Atherosclerosis 1979. **34**:419-429.

6.   Glomset, JA *J. Lip. Res.* 1968. 9:155-163

7.   Assmann, G, Schmitz, G and Brewer, HB. in Scriver, CR, Beaudet, AL, Sly, WS and Valle, D (eds) *The Metabolic Basis of Inherited Disease.* 6th edition. McGraw-Hill Information Services. New York. 1989. pp 1267-1282.

8.   Breslow, JL. in Scriver, CR, Beaudet, AL, Sly, WS and Valle, D (eds) *The Metabolic Basis of Inherited Disease.* 6th edition. McGraw-Hill Information Services. New York. 1989. pp 1251-1266.

9.   Assmann, G, von Eckardstein, A, and Funke, H. Current opinion in lipidology. 1991. 2: in press

10.   Norum, R.A., J.B. Lakier, S. Goldstein, A. Angel, R.B. Goldberg, W.D. Block, D.K. Nofze, P.J. Dolphin, J. Edelglass, D.D. Bogorad, and P. Alaupovic *N. Engl. J. Med.* 1982. **306**:1513-1519

11.   Schaefer, E.J., J. Ordovas, S. Law, G. Ghiselli, L. Kashyap, L. Srivastava, W.H. Heaton, J. Albers, W. Connor, Lindgren, F., A. Lemeshev, J. Segrest, and H.B. Brewer Jr. *J. Lip. Res.* 1984. 26:1089-1101.

12.   Karathanasis, S.K., E. Ferris, and I.A. Haddad. *Proc. Natl. Acad. Sci. U.S.A.* 1987. **84**:7198-7202.

13.   Ordovas, J.M., D.K. Cassidy, F. Civeira, C.L. Bisgaier and E.J. Schaefer. 1989. *J. Biol. Chem.* **264**:16339-16342.

14.   Deeb, SS, Cheung, MC, Peng, R, and Knopp, RH. *Circulation* 1990. **82** (Suppl. II): 424 (abstract)

15. Funke, H., A. von Eckardstein, P.H. Pritchard, M. Karas, J.J. Albers, and G. Assmann. 1991. *J. Clin. Invest. 1991, 87:375-380*

16. Schmitz, G., Lackner, K., and Robenek, H. *J. Clin. Invest. 1991, 87: in press*

17. von Eckardstein, A, Funke, H, Römling, R, Motti, C, Sandkamp, M, Noseda, Fragiacomo, and Assmann, G. *unpublished*

18. Forte, TM Nichols, AV, Krauss, RM, and Norum, RA. *J. Clin. Invest.* 1984. **74**:1601-1613

19. von Eckardstein, A., H. Funke, M. Walter, K. Altland, A. Benninghoven, and G. Assmann. *J. Biol. Chem.* 1990. **265**:8610-8617.

20. Nichols, W.C., F. Dwulet, J. Liepnieks, and M.D. Benson. 1988 *Biochem. Biophys. Res. Comm.* **156**, 762-768

21. Nichols, WC, Gregg, RE, Brewer, HB, and Benson, MD. *Genomics* 1990. 8:313-323.

22. Franceschini, G., Baio, M., Calabresi, L, Sirtori, CS, and Cheung, MC. *Biochim. Biophys. A.* 1990, 1043:1-6.

23. Franceschini, G., Calabresi, L, Tosi, C., Gianfranceschi, G., Sirtori, CR, and Nichols, AV. *J. Biol. Chem.* 1990. **265**:12224-12231.

24. von Eckardstein, A., H. Funke, A. Henke, K. Altland, A. Benninghoven, and G. Assmann, G. *J. Clin. Invest.* 1989. **84**:1722-1730.

25. Utermann, G., Haas, J., Steinmetz, A., Paetzold, R., Rall, S.C., and Weisgraber, K.H. (1984): *Eur. J. Biochem.* 144:326-331.

26. Jonas, A., A. von Eckardstein, K.E. Kezdy, A. Steinmetz and G. Assmann. 1991. *J. Lipid Res.* 1991. **32**: *in press*

27. Rall, S.C., K.H. Weisgraber, R.W. Mahley, Y. Ogawa, C.J. Fielding, G. Utermann, J. Haas, A. Steinmetz, H.J. Menzel, and G. Assmann *J. Biol. Chem.* 1984. **259**:10063-10070.

28. Ponsin, G., A.M. Gotto, G. Utermann, and H.J. Pownall. *Biochem. Biophys. Res. Comm.* 1985. **133**:856-862.

29. 72. Utermann, G, Steinmetz, A, Paetzold, R, Wilk, J, Feussner, G, Kaffearnik, H, Müller-Eckhardt, C, Seidel, D, Vogelberg, KA, and Zimmer, F. *Hum. Genet.* 1982. **61**:329-337.

30. A-II deficiency

31. Lohse, P., M.R: Kindt, D.J. Rader, and H.B. Brewer. *J. Biol. Chem.* 1990. **265**:10061-10064.

32. Menzel, H.J., E. Boerwinkle, S. Schragl-Will and G. Utermann. *Hum. Genet.* 1988. **79**:368-372.

33. Menzel, H.J., Sigurdsson, G. Boerwinkle, E. Schrangl-Will, S., Dieplinger, H., and G. Utermann. *Hum. Genet.* 1990. 84:344-346.

34. R.B. Weinberg, M.K. Jordan, and A. Steinmetz. *J. Biol. Chem.* 1990. **265**:18372-18378.

35. Boerwinkle, E., S. Visvikis, and L. Chan. *Nucleic Acids Res.* 1990. **18**:4966.

36. von Eckardstein, A, Funke, H, Schulte, M, and Assmann, *manuscript in preparation*

37. von Eckardstein, A., Holz, H., Sandkamp, M, Weng, W, Funke, H, and Assmann, G., *J. Clin. Invest.* 1991, in press

38. Norum K.R., Gjone E. & Glomset J.A. in The Metabolic Basis of Inherited Disease (eds. Scriver C.R. et al.) 1181-1194 (1989).

39. Taramelli, R, Pontoglio, M, Candini, G, Ottolenghi, S, Dieplinger, H, Catapano, A, Albers, J, Vergani, C and McLean, J. *Hum. Genet.* 1990, 85*:195-199*

40. Funke, H, von Eckardstein, A, Hornby, K, Hayden, M, Pritchard, PH, Albers, JJ, Frohlich, J, Gerdes, Faergeman, O, Dachet, J, Jacotot, B, and Assmann, G. *manuscript in preparation*

41. Funke, H, von Eckardstein, A, Pritchard, PH, Kastelein, JJP, Droste, C, Albers, JJ, and Assmann, G. *Proc. Natl. Acad Sci. U.S.A.* 1991, in press

42. Brown, ML, Hesler, C, and Tall, AR. *Current opinion in lipidology.* 1990. 1:123-127.

43. Brown, ML, Inazu, A, Hesler, CB, Agellon, LB, Mann, C, Whitlock, ME, Marcel, YL, Milne, RW, Koizumi, J, Mabuchi, H, Takeda, R and Tall AR. *Nature* 1989. 448-451.

44. N. Engl. J. Med.

APOLIPOPROTEIN E POLYMORPHISMS AND THE GENETIC

HETEROGENEITY OF FAMILIAL DYSBETALIPOPROTEINEMIA

Louis M. Havekes[1], Arn M.J.M. van den Maagdenberg[2], Peter de Knijff[1], Monique Mulder[1] and Rune R. Frants[2]

[1]Gaubius Institute TNO, P.O. Box 612, 2300 AP Leiden
[2]Dept. of Human Genetics, 2300 RA Leiden, The Netherlands

INTRODUCTION

Chylomicrons are secreted from the intestine into the lymph, whereafter they enter the bloodstream. In the bloodstream these particles are partly lipolyzed by lipoprotein lipase (LPL) and the chylomicron-remnant particles just formed are rapidly taken up by hepatic lipoprotein receptors. The liver secretes VLDL particles and, similar to chylomicrons in the bloodstream, VLDLs are partly lipolyzed by LPL and the VLDL-remnants or IDL particles formed are taken up with high affinity by the hepatic LDL receptors.

In patients with type III hyperlipoproteinemia or Familial Dysbeta-lipoproteinemia (FD) the uptake of chylomicron- and VLDL-remnants is impaired, leading to increased plasma cholesterol and triglyceride levels concomitant with increased risk for atherosclerosis (1). The uptake of the remnants is mediated by apolipoprotein (apo)E which represents one of the major protein constituents of these particles.

Apo E is a polymorphic protein consisting of three major isoforms (E2, E3 and E4) separated upon isoelectric focusing and coded for by three codominant alleles at one single gene locus. E3 is commonly assumed to be the wild type form, E2 is derived from E3 by an arg → cys substitution at position 158 (E2[arg158 → cys]), whereas the common E4 is designated E4[cys112 → arg].

In contrast to E4 and E3, E2 does not bind efficiently to hepatic lipoprotein receptors. E2E2 homozygosity, therefore, leads to an impaired clearance of remnants and is therefore the underlying metabolic defect of FD. In the general Caucasian population the frequency of the E2E2 phenotype is 0.7 to 1.0%. Although almost all E2E2 homozygotes express the impaired clearance of remnants, only about 1 to 4% of all E2E2 homozygotes develop type III hyperlipoproteinemia or FD at later age (2). Obviously, in case of E2E2 homozygosity, FD is a genetic disease that inherits recessively and that needs additional hyperlipoproteinemic factors to be expressed (multifactorial disease). It is assumed that these factors may be environmental factors, like nutrition, obesitas, and hormonal status (insulin, thyroid, sex hormones), or genetic factors. These additional genetic factors could

DNA Polymorphisms as Disease Markers, Edited by D.J. Galton and
G. Asmann, Plenum Press, New York, 1991

71

be (unknown) genes causing hyperlipoproteinemia or additional heterogeneity in the E*2 allele.

RESULTS

The aim of our study is to evaluate additional heterogeneity in the APOE gene in relation to the prevalence of FD. For this, we used the following techniques: a) isoelectric focusing with and without prior modification with cysteamine followed by immunoblotting. This gives the number of cysteine residues present in the E protein, b) screening for known mutations in the APOE gene by performing polymerase chain reaction (PCR) amplification of genomic DNA corresponding to the 5' part of exon 4. In this part of the gene all FD-causing apo E mutations described thusfar have been localised. After PCR we performed hybridisation with allele-specific synthesised oligonucleotides (ASO) or digestion with allele-specific restriction enzymes followed by polyacrylamide gelelectrophoresis (ASRE).

With a combination of the techniques of ASO and ASRE and that of isoelectric focusing with and without cysteamine modification, we were able to show that all sixty E2E2 homozygous FD patients were homozygous for the mutation E2[arg158 → cys] (Table 1). The presence of other known mutations could not be detected using ASRE and ASO, implying that in these E2E2 homozygous patients the expression of FD is most probably not due to mutations in the APOE gene in addition to the arg158 → cys mutation. In controls also all E*2 alleles tested (n=176) exhibit the arg158 → cys mutation, except two. These latter two are presently being characterised by DNA sequencing.

Besides 60 E2E2 homozygous FD patients, we also found FD patients without homozygosity for the arg158 → cys mutation. Six FD patients exhibited the E3E2 phenotype (Table 2). Upon cysteamine modification and isoelectric focusing, we found that in these patients the E2 isoform contains only one cysteine residue. Sequencing of the 5' part of exon 4 learned that the E*2 allele codes for the E2[lys146 → gln] variant. We designed synthetic oligonucleotides (19-mers) against this mutation and could show, using the ASO technique, that all six FD patients with the E3E2 phenotype carry this rare allele.

Family studies showed that all subjects carrying this allele expressed FD except two subjects; one with a very low body mass index, the other with a young age. This implies that in case of the E2[lys146 → gln] variant FD is transmitted in a dominant manner with high penetrance (3,4). Genealogical studies will be performed in order to evaluate whether these families share common ancestry.

Table 1
APOE genotyping of FD patients with E2E2 phenotype and of controls

subjects	phenotype	number	genotype
FD patients	E2E2	60	E2[arg158→cys] E2[arg158→cys]
Controls	E2E2	11	E2[arg158→cys] E2[arg158→cys]
	E2E2	2	E2[arg158→cys] E2[?]
	E2E3	150	E2[arg158→cys] E3

Table 2
FD patients without E2E2 homozygosity

Phenotype	number	characteristics	genotype
E2E3	6	E2 contains one cys residue	E2[lys146→gln] E3
E3E3	5	part of E3 contains no cys residues	E3-Leiden E3

Among our population of FD patients five exhibit the E3E3 homozygous phenotype as determined by isoelectric focusing (Table 2). After cysteamine treatment only a minor part of the E3 band moves to the E4 band. This suggests that all these five, apparently unrelated patients, display heterozygosity for the E3-Leiden variant as previously described for one of these patients (5). From these five subjects we amplified both exon 3 and exon 4 (5' and 3' part, separately). After electrophoresis and ethidium bromide staining we found a double band for the 5' part of exon 4. Further analyses using hybridisation with a APOE cDNA probe and treatment with PstI evaluated the presence of an insertion of about 20 bp in the 5' part of exon 4.

After cloning in m13 and sequencing, we found that the E*3-Leiden allele is derived from the E*4 allele by partial gene duplication, encompassing 21 nucleotides (codons 121-127 or 120-126), resulting in a tandem repeat of 7 amino acids (6,7). We designed synthetic oligonucleotides (21-mers) against the junction region and used the ASO and ASRE techniques to screen family members for carrying the E*3-Leiden allele.

Genealogical studies showed that all five E*3-Leiden probands appeared to share common ancestry between the year 1650 and 1700 (Fig.1). In total 42 E*3-Leiden allele carriers were found among 128 family members. All family members carrying the E*3-Leiden allele appeared to exhibit lipoprotein parameters characteristic for FD, although with a wide range in severity.

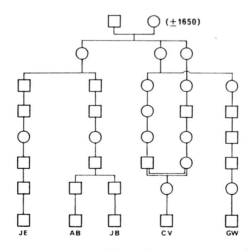

Figure 1.
Common ancestry for the five E3-Leiden probandi.

There is some overlap between E*3-Leiden allele carriers and non-carriers with respect to IDL-cholesterol and, although less pronounced, with respect to the ratio (VLDL + IDL)-cholesterol (d < 1.019 g/ml)/plasma triglyceride levels (Fig.2A,B). The plasma APOE levels are high in the E*3-Leiden allele carriers and show no overlap with that of the non-carriers (Fig.2C).

We conclude that all E*3-Leiden allele carriers show lipid parameters characteristic for FD, and thus, the E3-Leiden variant behaves like a dominant trait in the expression of FD.

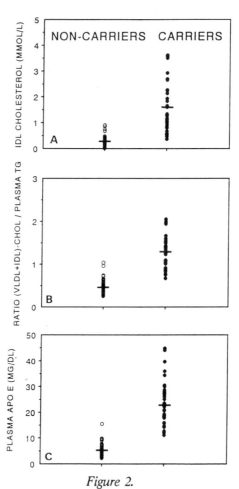

Figure 2.
Plasma lipoprotein parameters of E*3-Leiden allele carriers and non-carriers.

DISCUSSION

From these results it is obvious that FD is a genetic heterogeneous disease. In case of the E2[arg158 → cys] variant, FD is a recessively inherited multifactorial disease, whereas in case of the rare apoE variants (E2[lys146 → gln] and E3-Leiden) FD is dominantly inherited with a high rate of penetrance. The reason for this difference in mode of inheritance of FD between the common E2[arg158 → cys] variant and the rare E2[lys146 → gln] and E3-Leiden variants is at present subject to speculation.

Innerarity et al. (8) showed that after cleavage the E2[arg158 → cys] variant by thrombin in a N-terminal 22 kD fragment and a 12 kD C-terminal peptide, the receptor binding activity of the 22 kD N-terminal fragment is restored. The same occurs with the 22 kD fragment of the E3-Leiden variant (7). This suggests that both the cystein residue at position 158 as well as the two extra α-helices in apoE3-Leiden formed by the tandem repeat at residues 120-126 change the conformation of the apoE molecule in such a way that the C-terminal part of the protein prevents an efficient interaction of the binding domain with the receptor. Innerarity et al. (9) also showed that βVLDL from E2E2 homozygotes restores its binding activity after body weight reduction concomitant with a dramatic fall in plasma cholesterol and triglycerides. This suggests that the binding activity of E2 depends on the lipid composition of the βVLDL particle. Preliminary results show that βVLDL from hyperlipidemic E3-Leiden subjects is less active in binding than βVLDL from normolipidemic E3-Leiden subjects, also suggesting that the binding activity of E3-Leiden depends on the lipid composition of the VLDL particle, although less dramatic than that of the E2 variant.

Taken these facts of modulation the receptor binding activity of both apoE variants together, we conclude that both E2[arg158 → cys] as well as E3-Leiden need additional factors, like lipid composition of the VLDL particle to express FD. However, almost all E3-Leiden heterozygotes develop FD, whereas in case of E2[arg158 → cys] homozygosity FD is developed in only about 4% of all E2E2 homozygotes. Obviously, the two additional α-helices in the E3-Leiden variant (residues 120-126) affect the binding activity more or less irreversibly, whereas the arginine to cysteine substitution at position 158 exhibits a highly reversible effect on the binding activity.

In case of the rare apoE (E2[lys146 → gln]; E3-Leiden), FD is dominantly inherited with a high rate of penetrance, whereas in case of the E2[arg158 → cys] variant FD behaves like a recessively inherited multifactorial desease. The reason for the dominant mode of inheritance of FD in subjects with the rare apoE variants is currently under investigation. Preliminary studies showed that in case of the E2[lys146 → gln] variant, the VLDL particles are defective as substrate for lipoprotein lipase (LPL), resulting in βVLDL particles with high relative amounts of triglyceride. As VLDL remnants with relatively low cholesteroi/triglyceride ratios are less efficient in binding of the LDL receptor, as compared to remnants with higher ratios, we suggest that this defect as substrate for LPL might explain the dominant mode of inheritance of FD in E2[lys146 → gln] allele carriers.

The VLDL particles from subjects heterozygous for the E3-Leiden variant are normally lipolysed by LPL. The question remains therefore, why are these particles defective in binding to the receptor, irrespective the presence of normal apoE3 molecules. We studied the distribution of apoE3-Leiden and common apoE3 on the different lipoprotein fractions. Using isoelectric focusing with and without cysteamine modification, we were able to show that the apoE of VLDL and IDL density fractions mainly consisted of E3-Leiden, whereas the HDL fraction was relatively enriched in the common E3 variant. Although we did not study yet the reason for this difference in distribution, our observations could explain the dominant mode of inheritance of FD in case of E3-Leiden allele carriers as the VLDL particles on these subjects are phenotypically "almost" homozygous for the apoE3-Leiden variant.

In conclusion, FD is a genetically heterogeneous disease. It displays a recessively inherited multifactorial disease in case of the common E2[arg158 → cys] variant, and a dominantly inherited disease with a high rate of penetrance in case of rare apoE variants. Also in the literature, a number of rare apoE variants have been described that result in dominantly inherited FD (4,10,11). From a clinical point of view, we recommend that all patients with elevated plasma cholesterol and triglyceride levels concomitant with increased cholesterol/triglyceride ratios in the VLDL fraction should be analysed for apoE phenotype and/or genotype. In case of E2E2 homozygosity this analysis will only sustain the FD diagnosis. However, when the suggested FD patients do not exhibit the common E2E2 homozygosity, the patients might carry a rare APOE allele, and thus family studies are indicated as in these cases early diagnosis of FD is feasible.

Following this strategy of FD diagnosis, we found that about 20% of all patients diagnosed as FD in our lipid clinics based on lipid levels exhibit non-E2E2 homozygosity. After performing family studies of both E2E2 FD patients as well as part of non-E2E2 homozygous FD patients, we eventually found that about 50% of all patients exhibit heterozygosity for rare apoE variants.

ACKNOWLEDGEMENTS

This study was financially supported by the Nederlandse Hartstichting (A. van de M. #88.086 and M. M. #87.025) and Praeventiefonds (P. de K. #28-1716).

REFERENCES

1 Mahley, R. W., and S. C. Rall. 1989. Type III hyperlipoproteinemia (dysbetalipoproteinemia): The role of apolipoprotein E in normal and abnormal lipoprotein metabolism. In The Metabolic Basis of Inherited Disease. C. R. Scriver, A. L. Beaudet, W. S. Sly, and D. Valle, editors. 6th ed. McGraw-Hill Book Co., New York. 1195-1213.

2 Utermann, G., K. H. Vogelberg, A. Steinmetz, W. Schoenborn, N. Pruim, M. Jaeschke, M. Hees, and H. Canzler. 1979. Polymorphism of apolipoprotein E. Genetics of hyperlipoproteinemia type III. Clin. Genet. 15: 37-62.

3 Smit, M., P. de Knijff, R. R. Frants, E. C. Klasen, and L. M. Havekes. 1987. Familial dysbetalipoproteinemic subjects with the E3/E2 phenotype exhibit an E2 isoform with only one cysteine residue. Clin. Genet. 32:335-341.

4 Smit, M., P. de Knijff, E. van der Kooij-Meijs, C. Groenendijk, A. M. J. M. van den Maagdenberg, J. A. Gevers Leuven, A. F. H. Stalenhoef, P. M. J. Stuyt, R. R. Frants, and L. M. Havekes. 1990. Genetic heterogeneity in familial dysbetalipoproteinemia. The E2(Lys146→Gln) variant results in a dominant mode of inheritance. J. Lipid Res. 31:45-53.

5 Havekes, L., E. de Wit, J. Gevers Leuven, E. Klasen, G. Utermann, W. Weber, and U. Beisiegel. 1986. Apolipoprotein E3-Leiden. A new variant of human apolipoprotein E associated with familial type III hyperlipoproteinemia. Hum. Genet. 73:157-163.

6 van den Maagdenberg, A. M. J. M., P. de Knijff, A. F. H. Stalenhoef, J. A. Gevers Leuven, L. M. Havekes, and R. R. Frants. 1989. Apolipoprotein E*3-Leiden allele results from a partial gene duplication in exon 4. Biochem. Biophys. Res. Comm. 165:851-857.

7 Wardell, M. R., K. H. Weisgraber, L. M. Havekes, and S. C. Rall, Jr.. 1989. Apolipoprotein E3-Leiden contains a seven-amino acid insertion that is a tandem repeat of residues 121-127. J. Biol. Chem. 264:21205-21210.

8   Innerarity, T. L., K. H. Weisgraber, K. S. Arnold, S. C. Rall, Jr., and R. W. Mahley. 1984.   Normalization of receptor binding of apolipoprotein E2. Evidence for modulation of the binding site conformation. J. Biol. Chem. 259:7261-7267.

9   Innerarity, T. L., Hui, D. Y., Bersot, T. P., and Mahley, R. W. 1886. Type III hyperlipoproteinemia: a focus on lipoprotein receptor-apolipoprotein E2 interactions. In Lipoprotein Deficiency Syndroms. A. Angelin, And J. Frohlich, editors. Plenum Publishing Co., New York. 273-288.

10 Rall, S. C., Jr., Y. M. Newhouse, H. R. G. Clarke, K. H. Weisgraber, B. J. McCarthy, R. W. Mahley, and T. Bersot. 1989. Type III hyperlipoproteinemia associated with apolipoprotein E phenotype E3/3. Structure and genetics of an apolipoprotein E3 variant. J. Clin. Invest. 83. 1095-1101.

11 Mann, W. A., R. E. Gregg, R. Ronan, F. Thomas, L. A. Zech, and H. B. Brewer, Jr. 1988. Apolipoprotein E1-Harrisburg: point mutation resulting in dominant expression of type III hyperlipoproteinemia. 1988. Arteriosclerosis. 8:612a (Abstr.)

ABNORMALITIES OF APOLIPOPROTEIN B METABOLISM

IN THE LIPID CLINIC

Alberto Corsini, Alberico L. Catapano, Maria Mazzotti,
Guido Franceschini, Cesare Romano*, Remo Fumagalli,
and Cesare R. Sirtori

Institute of Pharmacological Sciences, University of
Milan, and *G.Gaslini Institute, First Dept.
Pediatrics, University of Genoa, Italy

Abstract: Apolipoprotein (apo) B-100 is the main protein
component of low density lipoprotein (LDL) and
plays a crucial role in cholesterol and
lipoprotein metabolism. Elevated LDL cholesterol
concentrations may derive either from a
defective clearance of LDL or from a reduced
removal. In the present report two apo B
abnormalities that cause hypercholesterolemia
were investigated. The case of a family affected
with defective apo B-100 and of a woman with
severe hypercholesterolemia and hyperapobetalipoproteinemia are presented.

INTRODUCTION

Apolipoprotein (apo B), the main protein constituent of low
density lipoprotein (LDL), very low density lipoprotein
(VLDL) and chylomicrons, plays a pivotal role in lipoprotein
metabolism (1). Apo B occurs naturally in two closely
related forms (apo B-100 and B-48), encoded by the same gene
(2). Apo B-100 is synthesized by the liver and is required
for the assembly and secretion of VLDL. Apo B-48 is
associated mainly with chylomicrons, chylomicron remnants
and, in humans, it is believed to be synthesized solely by
the intestine (1). The complete aminoacid (aa) sequence of
human apo B was recently deduced from cDNA (3,4). Apo B-100
consists of a 4536 aa single polypeptide chain, apo B-48
represents the amino-terminal 47% of apo B. In humans apo B-
100 is the predominant, if not exclusive, protein
constituent of LDL, a lipoprotein that carries about 70% of
plasma cholesterol, and is responsible for the receptor-
mediated catabolism of LDL (5). The importance of LDL apo B-
100/LDL receptor interaction in maintaining cholesterol
homeostasis is best illustrated by familial hypercholesterolemia (FH), a genetic disease in which a defective

LDL receptor expression triggers an increase of plasma LDL cholesterol levels and premature coronary heart disease (CHD) (5).

Most individuals with hypercholesterolemia, however, possess a normal number of LDL receptors (6). Therefore, in most hypercholesterolemic patients other factors must contribute to the increased plasma levels of LDL, such as defective clearance or increased production.

In principle, elevated LDL concentrations that result from an inefficient clearance of LDL, may derive either from receptor or ligand defects. Genetic abnormalities of apo B-100, resulting in a decreased "in vitro" binding of LDL to their receptor, and hence a reduced "in vivo" catabolism of LDL, have been recently identified in some hypercholesterolemic patients (7-8). This genetic disease has been named familial defective apo B-100 (FDB) (7-9). In the families described so far, a single substitution at base 11039 (G to A) leads to an arginine to glutamine change in a mature protein at aa 3500. An overproduction of LDL may derive either from an increased synthesis of apo B-containing lipoproteins or from a decreased uptake of VLDL remnants thus allowing for a more efficient conversion of VLDL to LDL (6). Cases of this type or of with similar clinical and biochemical pattern may be frequently encountered in the activity of a lipid clinic.

In this report, we focus on the case of a family affected with FDB and on that of a middle age woman with a significant overproduction of apo B-containing lipoproteins.

SUBJECTS AND METHODS

Subjects

The cases of a family and of a woman presenting with hypercholesterolemia were investigated.

I) P. family

The proband (P.D.) was referred to us because of hypercholesterolemia at age of 10 yr (298 mg/dl); all other laboratory tests, including plasma triglycerides, glucose, IRI and T_3 were within normal range. He had no xanthomas, xanthelasmas or thickening of the Achilles tendon. He is currently aged 16, in good health and consuming a low fat, low cholesterol diet. Upon screening of the family members, hypercholesterolemia was detected in the proband's father (P.V.), grandfather (P.At.), great-uncle (P.Al.) and great-aunt (P.W.) (Fig. 1).

II) Patient S.L.

The patient (S.L.), a 52 year old woman, was referred to us in September 1987, because of marked hypercholesterolemia with extensive xanthomata at the elbows, Achilles tendons, hands and fingers

At the first examination, the patient had a serum total cholesterol concentration of 658 mg/dl. Coronary arteriography showed complete occlusion of the right coronary, and severe stenosis of the left main (80%), anterior descending (90%) and circumflex (60%). In spite of the clearcut features of FH, elevated serum cholesterol or coronary artery disease were not reported in any of the first degree relatives.

Since October 1988, the pt has been treated with a

combination of simvastatin (40 mg/day) and cholestyramine (12 g/day), with periodic LDL-aphereses. This therapy has proven effective, maintaining cholesterol concentrations around 260 mg/dl.

Plasma lipids and lipoproteins

Plasma lipids and HDL cholesterol were determined by standard ultracentrifugal/precipitation methods.
Apolipoprotein A-I and B were determined using a commercially available radial immunodiffusion assay.
LDL were isolated from plasma by ultracentrifugation at d 1.019-1.063 g/ml. Lipoprotein-deficient serum (LPDS) was obtained as the serum fraction at d > 1.25 g/ml.
All lipoproteins and LPDS were extensively dialyzed and stored at 4°C and at -20°C respectively after sterilization. LDL were labeled with ^{131}I or ^{125}I to a specific activity of 150-230 cpm/ng, according to Bilheimer et al. (10).

"In vitro" studies

Tissue culture experiments were performed using fibroblasts obtained from patients and normolipidemic subjects. Cells were cultured as previously reported (11). The ability of LDL from P. family members and controls to bind the LDL receptor was determined in an "in vitro" competition binding assay using control ^{125}I-labeled LDL as the ligand. Uptake and degradation of ^{125}I-LDL by human skin fibroblasts (HSF) were determined as previously described (11).
The ability of LPDS to interfere with the LDL-receptor interaction was investigated after preincubation of cells for 2h at 37°C in a medium with increasing concentration of LPDS from patients S.L. and from a control. To determine the amount of cholesteryl |[14]C|oleate formation, normal human fibroblasts were preincubated for 25h at 37°C in 10% LPDS. The medium was then changed with a fresh one containing 0.2 mM |[14]C|oleate/albumin in the absence or presence of various concentrations of LDL. Cholesteryl |[14]C|oleate was isolated by TCL and measured by liquid scintillation counting, using |[3]H|cholesteryl oleate as the internal standard (12).

Turnover of autologous and homologous labeled LDL

In order to verify "in vivo" an abnormality of LDL metabolism, patient S.L. received homologous and autologous LDL, respectively labeled with ^{125}I and ^{131}I, essentially according to the protocol described by Grundy et al. (13). The plasma decay curve for both isotopes was monitored by collecting plasma samples at 10 and 20 min after injection and at 1, 4, 8, 12 and 24h; thereafter samples were drawn every 12h for the subsequent 3 days, and finally, every 24h from the 4th throughout the 12th day. Determinations of radioactivity were made at each time interval on isopropanol-precipitated samples and plasma lipid, lipoprotein cholesterol distribution and LDL apo B levels were measured twice weekly.
The fraction catabolic rate (FCR) and transport rate were determined as described by Matthews (14).

RESULTS

Two apo B abnormalities that cause hypercholesterolemia were investigated: a) defective clearance of LDL, b) overproduction of LDL.

a) Reduced clearence of LDL

Familial defective apo B-100

In this study we investigated a family affected by primary moderate hypercholesterolemia (250-350 mg/dl). Eight members of the P. family were investigated (Fig. 1). In all affected members but one (P.E.) the phenotype was IIa.

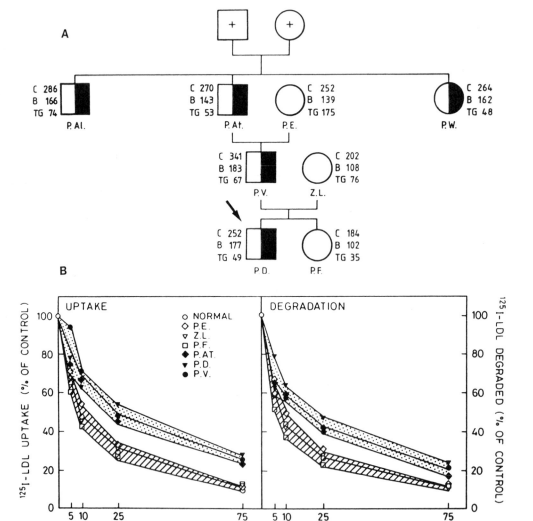

Fig.1 A) Pedigree of the P. family. The values of plasma cholesterol (C), plasma triglycerides (TG), and apolipoprotein B (B) are expressed as mg/dl.
B) Ability of LDL from the P. family members and from normal subjects to compete with normal ^{125}I-LDL for the uptake and degradation by human skin fibroblasts.

P.E. is a diabetic on insulin. The pattern of inheritance of the disease, hypercholesterolemia at young age, and CHD in one family member (P.V.) in the third decade of life are consistent with the diagnosis of heterozygous FH. Biochemical studies on cultured skin fibroblasts ruled out the classical FH (receptor deficiency). We then investigated the interaction between LDL from all family members or from normolipemic controls with the LDL receptor by an "in vitro" cell binding assay. LDL from the affected members displayed a reduced affinity (2.5 fold) for its receptor (Fig. 1). Invariably binding-defective LDL were associated with hypercholesterolemia in the P. family.

The average IC_{50} (concentration of LDL recquired for 50% reduction of the uptake and degradation of control ^{125}I-LDL by HSF) from affected vs unaffected members and controls were 21 $\pm$ 3 vs 10 $\pm$ 2 ug/ml and 16 $\pm$ 4 vs 8 $\pm$ 2 ug/ml for uptake and degradation, respectively. The incorporation of $|^{14}C|$oleate into cellular cholesteryl ester was used as a measure of lipoprotein delivery of cholesterol to cells. As shown in Table I LDL from the affected members were less effective than normal LDL in promoting cholesteryl $|^{14}C|$oleate formation, thus confirming that apo B-100 defective LDL deliver less biologically active cholesterol to the cell. The LDL from the affected members had a normal electrophoretic mobility, size and chemical composition. Analysis of DNA from family members showed a point mutation of the apo B gene leading to an Arg to Gln substitution at

TABLE I

ABILITY OF FAMILY P. MEMBERS LDL TO INDUCE CHOLESTERYL
ESTERIFICATION IN NORMAL HUMAN FIBROBLASTS

Family member	Cholesterol esterification rate (pmol/h per mg cell protein)	% of Control
Control	3480 $\pm$ 123.0	100
Affected		
P.A.	2934 $\pm$ 82.0	78
P.At.	2114 $\pm$ 9.3	53
P.W.	2562 $\pm$ 136.6	66
P.V.	1901 $\pm$ 145.1	46
P.D.	2638 $\pm$ 105.6	64
Unaffected		
P.E.	3071 $\pm$ 0.3	82
Z.L.	3335 $\pm$ 0.6	91
P.F.	3365 $\pm$ 217.0	92

Experimental conditions as described in Methods. Data are the average of triplicates $\pm$ SD aminoacid 3500 of the mature protein, that segregates with hypercholesterolemia and LDL defective binding (15). We conclude that our family is affected by a familial defective apolipoprotein B-100. This disorder was transmitted over three generations as an autosomal codominant trait. All affected members are heterozygotes.

b) Overproduction of LDL

Increased LDL synthesis

Table II shows the plasma lipid and apoprotein levels in patient S.L.. Her phenotype was IIa. Compared to control subjects, the pt showed a 2.5 fold increase of LDL mass, with a slight reduction of HDL. VLDL were cholesterol enriched and TG depleted; LDL and HDL composition was grossly normal, except for an increase in the LDL protein content. Plasma and LDL apo B levels were markedly elevated, the cholesterol/apo B ratio in LDL being in the range reported for patients with hyperapobetalipoproteinemia (16). Family studies showed, however, that the first degree relatives were normocholesterolemic, thus ruling out the diagnosis of FH. In fact studies on cultured skin fibroblasts obtained from patients S.L. showed that the uptake and degradation of LDL was about 75% of normal. The ability of LDL from the patient to compete with normal ^{125}I-LDL for uptake and degradation by HSF was also determined. S.L. LDL were as effective as control LDL in competing with normal ^{125}I-LDL for receptor activity (Fig. 2). The effect of control and pt's LPDS on the uptake and degradation of ^{125}I-LDL by HSF was also investigated. LPDS from the pt did not affect this process, thus excluding the presence in the pt's serum of factor/s (ie antibodies) capable of interfering with the LDL-receptor interaction. Thus the "in vitro" studies on the LDL receptor function fully support the conclusion that the LDL receptor interaction is normal in this patient.

TABLE II

PLASMA LIPID/LIPOPROTEIN DISTRIBUTION AND COMPOSITION IN PATIENT S.L.*

	Plasma	VLDL	LDL	HDL
		mg/dl		
Total cholesterol	425.0	25.2	361.8	38.0
Triglycerides	114.6	57.7	265.5	11.3
Proteins	N.D.	18.7	301.4	121.7
Apolipoprotein A-I	97.0	N.D.	N.D.	91.3
Apolipoprotein B	294.0	N.D.	283.5	N.D.

* Sample drawn June 1988

The kinetics of disappearance of autologous [131]I- and homologous ^{125}I-labeled LDL, investigated in the patient, failed to show any difference in the behaviour of the two lipoproteins; the FCR for patient S.L. and control were also essentially the same (0.34/day vs 0.30/day respectively). The production rate of LDL apo B in the steady-state conditions was increased almost 3-fold above normal (36.8 mg/Kg/day vs 12.9 mg/Kg/day respectively).

In order to reveal possible alterations at the apo B gene, the regulatory region upstream to the 5' end of the apo B gene was analyzed.

Sequence of this area did not reveal any mutation compared to a control sequence (data not shown). Although the exact mechanisms could not be identified, hypercholesterolemia in patient S.L. could be solely explained by an increased LDL biosynthesis.

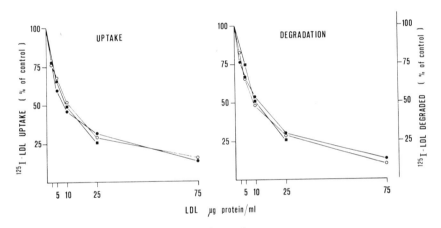

Fig. 2

Ability of LDL patient S.L. (●) and from two normolipidemic individuals (■ , ○) to compete with normal ^{125}I-LDL for receptors-mediated uptake and degradation by normal skin fibroblasts. The 100% control values for cellular uptake and degradation were 466 and 325 ng/mg of cellular protein, respectively.

DISCUSSION

Apo B-100 is the main apolipoprotein component of LDL, and is the major determinant for the lipoprotein metabolic pathway (1,5). Elevated LDL concentrations in plasma are caused by increased production of LDL or by reduced removal (6).

A reduced clearance of LDL may result from a defect either in the expression of LDL receptor, as in FH, or in apo B-

100, the physiological ligand. Familial defective apo B-100 is a genetic disorder presenting with moderate hypercholesterolemia and abnormal LDL that bind poorly to their receptors (7-9,17). This functional defect appears to be caused by a mutation in apo B-100 resulting in a delayed clearance of LDL from plasma. Three families with hypercholesterolemia and apo B defective LDL have been described. This condition segregates with a Gln for Arg substitution at aa 3500 of apolipoprotein B (9). We have recently reported a family in which moderate hypercholesterolemia segregates with binding defective LDL (17).

Biochemical studies on HSF ruled out the diagnosis of FH (receptor deficiency). LDL from the hypercholesterolemic family members are 2.5 fold less effective than normal in interacting with the LDL receptor. Apo B DNA analysis of the family members showed a point mutation at aa 3500, that segregates with hypercholesterolemia and binding defective LDL (15). We conclude that this family is affected by familial defective apo B-100. Four studies have examined the impact of this mutation in the population (18-21). Tybjaerg-Hansen (18) and Schuster (19) have showed that the frequency of the mutation was 3% in sujects with hypercholesterolemia. Innerarity et al. (21) estimated its frequency to be about 1/500 in the general population, ie similar to classic FH. In contrast Hamalainen et al. (20), upon screening of 552 hyperlipidemic subjects in Finland, failed to find a single case of FDB. It is possible, that mutations other than that at aa 3500 could hamper the binding of LDL to their receptor and lead to hypercholesterolemia.

FDB was first observed in subjects with moderate hypercholesterolemia. However, in the FDB patients reported in the U.K. and Scandinavian, studies cholesterol levels were comparable to FH. The family reported here has less severe hypercholesterolemia than those described by Schuster (19) and by Tybjaerg-Hansen (18) being similar to those reported by Innerarity (21).

The clinical features of FDB appear to be, at least in part, dependent upon the degree of hypercholesterolemia. Patients with moderate hypercholesterolemia do not display xanthomas and corneal arcus, although possessing all the other clinical features of FH (22) patients ie pattern of transmission, and hypercholesterolemia at an early age. On the contrary, patients with FDB and high plasma cholesterol resemble more closely FH patients (5,22). Therefore neither clinical nor routine laboratory tests distinguish this disorder from heterozygous FH and the optimal therapeutic approach required to achieve a satisfactory reduction of plasma cholesterol levels may also prove to be somewhat different. Preliminary data from our laboratory suggest that subject P.V. is poor responder to treatment with HMGCoA reductase inhibitors (23).

Another potential cause for elevated LDL (6) cholesterol concentrations is an enhanced input of LDL. This is particularly the case of a patient (S.L.) with all the clinical hallmarks of FH including elevated plasma cholesterol, xanthomas and early coronary and pheripheral arterial lesions but with a normal LDL receptor activity. In addition, LDL from patient S.L. possess a normal ability to

compete with ^{125}I-LDL for uptake and degradation by HSF. LPDS from this patient did not affect the LDL-receptor interaction, thus excluding the presence of interfering factor(s) (5). These "in vitro" studies clearly suggest that the patient is not affected by classical FH or FDB.
Hyperapobetalipoproteinemia (16), a pleiotropic phenotype characterized by LDL with a reduced cholesterol/apo B ratio, is associated with coronary artery disease. Patients affected by hyperapo B may be normolipidemic, hyper-triglyceridemic or hypercholesterolemic. HDL cholesterol and apo A-I levels are often low in plasma of these patients, and the increased number of LDL particles is due to an enhanced LDL production secondary to an augmented VLDL synthesis. Apo B/cholesterol ratio in the LDL from patient S.L. was increased and a turnover study performed using ^{125}I- homologous and ^{131}I autologous LDL, showed that the patient S.L. had a 3-fold increase in LDL-apo B synthesis, with a fractional catabolic rate within normal limits. The normal LDL receptor activity suggest that the increased LDL production observed in patient S.L. could be due to an enhanced production of apo B containing lipoproteins. Therefore the overall clinical picture of our patient S.L. clearly suggest a lipoprotein phenotype resembling hyperapobetalipoproteinemia.
These findings prompted us to investigate the presence of possible abnormalities in the promoter region of the apo B gene, theorically responsible for the down regulation of the hepatic synthesis of apo B-100. Studies at the 5' region of the apo B gene, however, failed to detect any alteration vs control published sequences. At present, therefore, it cannot be concluded what are the mechanisms responsible for the overproduction of LDL in the patient S.L., leading to severe CAD.
In summary, we have reported two clinical pictures, familial defective apo B-100 and overproduction of LDL, characterized by abnormalities of apo B-100 metabolism, leading to elevated LDL cholesterol levels. A more detailed understanding of the mechanism responsible for abnormalities of apo B-100 metabolism, leading to hypercholesterolemia will help to define a better strategy for treatment.

ACKNOWLEDGEMENTS

This work was partially supported by the C.N.R. Target Project on Biotechnology and Bioinstrumentation. Miss Maddalena Marazzini typed the manuscript.

REFERENCES

1) R.J. Havel and J.P. Kane, Structure and metabolism of plasma lipoproteins, in: "The Metabolic Basis of Inherited Disease", 6th Edition. C.R. Scriver, A.L. Beaudet, W.S. Sly, D. Valle, editors. McGraw-Hill, New York 1129 (1989).
2) L.M. Powell, S.C. Wallis, R.J. Pease, Y.H. Edwards, T.J. Knott, and J. Scott, A novel form of tissue-specific RNA processing produces apolipoprotein B-48 in intestine, Cell 50: 831 (1987).
3) T.J. Knott, R.J. Pease, L.M. Powell, S.C Wallis, S.C.

Rall, Jr., T.L. Innerarity, B. Blackhart, W.H. Taylor, Y. Marcel, R. Milne, D. Johnson, M. Fuller, A.J. Lusis, B.J. McCarthy, R.W. Mahley, B. Levy-Wilson, and J. Scott, Complete protein sequence and identification of structural domains of human apolipoprotein B, Nature 323: 734 (1986).

4) C-Y Yang, S-H Chen, S.H. Gianturco, W.A. Bradley, J.T. Sparrow, M. Tanimura, W-H Li, D.A. Sparrow, H. Deloof, M. Rosseneu, F-S Lee, Z-W Gu, A.M. Gotto, Jr., and L. Chan, Sequence, structure, receptor binding domain and internal repeats of human apolipoprotein B-100, Nature 323: 738 (1986).

5) J.L. Goldstein and M.S. Brown, Familial hyper-cholesterolemia, in: "The Metabolic Basis of Inherited Disease", 5th Edition. J.B. Standbury, J.B. Wyngaarden, D.S. Fredrickson, J.L. Goldstein and M.S. Brown editors. McGraw-Hill, New York 672 (1983).

6) S.M. Grundy, G.L. Vega, Causes of high blood cholesterol, Circulation 81: 412 (1990).

7) T.L. Innerarity, K.H. Weisgraber, K.S. Arnold, R.W. Mahley, R.M. Krauss, G.L. Vega, and S.M. Grundy, Familial defective apolipoprotein B-100: low density lipoproteins with abnormal receptor binding, Proc. Natl. Acad. Sci. 84: 6919 (1987).

8) G.L. Vega and S.M. Grundy, In vivo evidence for reduced binding of low density lipoproteins to receptors as a cause of primary moderate hypercholesterolemia, J. Clin, Invest. 78: 1410 (1986).

9) L.F. Soria, E.H. Ludwig, H.R.G. Clarke, G.L. Vega, S.M. Grundy, and B.J. McCarthy, Association between a specific apolipoprotein B mutation and familial defective apolipoprotein B-100, Proc. Natl Acad. Sci. 86:587 (1989).

10) D.W. Bilheimer, S. Eisenberg, and R.I. levy, The metabolism of very low density lipoprotein proteins. I. Preliminary in vitro and in vivo observations, Biochim. Biophys. Acta 260: 212 (1972).

11) A. Corsini, P. Roma, D. Sommariva, R. Fumagalli, and A.L. Catapano, Autoantibodies to the low density lipoprotein receptor in a subject affected by severe hypercholesterolemia, J. Clin. Invest. 78: 940 (1986).

12) J.L. Goldstein, S.K. Basu, and M.S. Brown, Receptor-mediated endocytosis of low density lipoprotein in cultured cells, Methods. Enzimol. 98: 241 (1983).

13) S.M. Grundy, G.L. Vega, and D.W. Bilheimer, Kinetic mechanisms determinating variability in low density lipoprotein levels and rise with age, Arteriosclerosis 5: 623 (1985).

14) C.M.E. Matthews, The theory of tracer experiments with ^{131}I-labelled plasma proteins, Physiol. Med. Biol. 2: 36 (1957).

15) A. Corsini, S. Fantappiè, L.F. Soria, B.J. McCarthy, A. Granata, F. Bernini, L. Romano, C. Romano, A.L. Catapano, and R. Fumagalli, Binding-defective low density lipoproteins in a family with primary hypercholesterolemia, in: 30th International Conference on the Biochemistry of Lipid, Abs C 14 (1989).

16) P.O. Jr. Kwiterovic, HyperapoB: a pleiotropic phenotype characterized by dense low density lipoproteins and associated with coronary artery disease, Clin, Chem. 34/8 (B): B71 (1988).

17) A. Corsini, S. Fantappiè, A. Granata, F. Bernini, A.L.

Catapano, R. Fumagalli, L.Romano, and C. Romano, Binding-defective low density lipoprotein in family with hypercholesterolaemia, <u>The Lancet</u> 1: 623 (1989).

18) A. Tybjaerg-Hansen, J. Gallagher, J. Vincent, R. Houlston, P. Talmud, A.M. Dunning, M. Seed, A. Hamsten, S.E. Humphries, and N.B. Myant, Familial defective apolipoprotein B-100 detection in the United Kingdom and Scandinavia and clinical characteristies of ten cases, <u>Atherosclerosis</u> 80: 235 (1990).

19) H. Schuster, G. Rahn, B. Kormann, J. Heppis, S.E. Humphries, L. Keller, G. Wolfram, and N. Zollner, Familial defective apolipoprotein B-100 in 18 cases detected in Munich, <u>Arteriosclerosis</u> 10: 517 (1990).

20) T. Hamalainen, A. Palotie, K. Aalto-Setala, K. Kontula, and M.J. Tikkannen, Absence of familial defective apolipoprotein B-100 in Finnish patients with elevated serum cholesterol, <u>Atherosclerosis</u> 82: 177 (1990).

21) T.L. Innerarity, Familial hypobetalipoproteinemia and familial defective apolipoprotein B-100: genetic disorders associated with apolipoprotein B, <u>Curr. Opin. Lip.</u> 1: 104 (1990).

22) J.L. Goldstein and M.S. Brown, Familial hypercholesterolemia, <u>in</u>: "Metabolic Basis of Inherited Disease", 6th Edition. C.R Scriver, A.C. Beanoht, W.S. Sly, D. Volle editors. McGraw-Hill, New York 1215 (1989).

23) A. Corsini, L. Romano, A. Granata, C. Romano, R. Fumagalli, and A.L. Catapano, Effect of simvastatin in subjects with binding-defective low density lipoprotein, Drugs Affecting Lipid Metabolism. In Press.

GENETIC VARIATION AT THE APOA-I, C-III, A-IV GENE COMPLEX: A CRITICAL REVIEW OF THE ASSOCIATIONS BETWEEN THE PSTI AND SSTI RFLPS AT THIS LOCUS WITH LIPID DISORDERS

Jose M. Ordovas, Fernando Civeira, Carmen Garces and Miguel Pocovi

Lipid Metabolism Laboratory. USDA Human Nutrition Research Center on Aging at Tufts University, Boston, USA, and Department of Cellular and Molecular Biology. University of Zaragoza, Spain

INTRODUCTION

Low levels of plasma high density lipoprotein (HDL) cholesterol constitute one of the major risk factor for coronary heart disease (CHD) in Western societies. The association of low HDL cholesterol levels with an increased risk of CHD was noted almost forty years ago and it has been supported by a large number of studies carried out primarily during the past 15 years [1-4]. While the genetic factors affecting low density lipoprotein (LDL) levels are relatively well known [5,6], there is very little knowledge about the factors involved in the genetic control of plasma HDL.

HDL particles are isolated by ultracentrifugation in the density region between 1.063 and 1.21 mg/dl. HDLs are very heterogeneous, with molecular weight ranging from 250,000 up to 400,000. This variation is in part due to their lipid to protein ratio, and in part to the protein composition. The major protein components of HDL are apolipoproteins (apo) A-I and apoA-II. Other minor components include apoA-IV, apoE, apoD and the C apolipoproteins. Genes coding for these proteins, together with others involved in lipoprotein metabolism, such as cholesteryl ester transfer protein (CETP), hepatic lipase (HL), lipoprotein lipase (LPL) may play a role in the genetic regulation of HDL levels. These genes have been isolated and characterized in the last few years, and probes have been derived allowing the study of genetic variation at this loci. The apoA-I gene locus at the 11q23 of chromosome 11 is in close linkage with the apoC-III and the apoA-IV genes [7,8]. Mutations within this locus have been associated with marked HDL cholesterol deficiency and severe premature atherosclerosis. Two such mutations have clearly been documented [9,10]; one resulting in a deletion of the entire apoA-I/C-III/A-IV gene complex [11] and a second mutation resulting in a DNA re-arrangement affecting the adjacent apoA-I and apoC-III genes [12]. At least 12 restriction fragment length polymorphisms (RFLPs) have been identified at this locus, and some of them have been studied extensively with regard to their usefulness as genetic markers for CHD and or genetic dyslipidemias. The first RFLP reported to be associated with a lipid disorder was the SstI RFLP due to single base mutation in the 3' untranslated region of the apoC-III gene [13]. In their original study, Galton et al. reported an strong association, in a British population, between the presence of the minor allele and hypertriglyceridemia [14]. Subsequent reports from this group and others have found similar association, as well as an association with increased risk of CHD. Considering the known inverse correlation between triglycerides and HDL cholesterol levels, is quite possible that an alteration on HDL metabolism could be responsible for such findings. The association between the SstI RFLP and hypertriglyceridemia has not been found in all populations. Moreover, studies in non Caucasian populations have demonstrated a high prevalence of the rare allele and no significant effect on lipid levels. In Caucasian populations, this RFLP is in linkage disequilibrium with two additional RFLPs in this locus, namely the MspI in the third intron of the apoA-I gene and the XbaI located 3' to the apoA-IV gene [15]. It is quite possible that differences in haplotypes could be responsible for the disparity observed among different studies. Another RFLP 300 bp 3' to the apoA-I gene, has been identified using PstI. In a previous report we found an association between the rare allele of this RFLP and premature CHD. In addition the frequency of the rare allele was also demonstrated to be significantly elevated in probands of kindreds with familial hypoalphalipoproteinemia [16]. Similar associations have been reported by other investigators. However most reports, including a more extensive study carried out by our group, have failed to confirm such effects [15]. More recently, Scott et al. have reported a linkage between the XmnI RFLP located 2.5 kb 5' to the apoA-I gene and familial combined hyperlipidemia (FCHL) [17]. This linkage had been previously suggested by Hayden et al. in an association study [18].

In the present report we review the present knowledge on the frequency distribution of the SstI and PstI RFLPs in different populations and their associations with lipid levels. In addition we are presenting original data regarding the frequency of these RFLPs in a Spanish population.

METHODS

Population

A group of males and females (n=229), factory workers in the region of Aragon (Spain), were randomly selected from an ongoing prospective study to characterize the major coronary risk factors in this population. The only reasons for exclusion were previous history of CHD or triglyceride levels above 350 mg/dl. A group of subjects (n = 35) with triglycerides above 350 mg/dl was also studied. All subjects were Caucasians, and their families have been residing in the region for at least three generations. The ethnic background of this population is a composite of Mediterranean and Nordic types.

Lipoprotein and apolipoprotein analysis

Blood was drawn after an overnight fast of at least 12 hrs in tubes containing 0.1% EDTA. Plasma cholesterol, triglyceride and HDL concentrations were determined using enzymatic assays. For HDL cholesterol determinations, VLDL and LDL were precipitated using Dextran sulfate-Mg [19].

DNA isolation and genomic blotting analysis

Genomic DNA was isolated from lymphocytes as previously described [20]. Seven ug of DNA were digested for at least three hours with either PstI or SstI using the buffers provided by the manufacturer and a 10 fold excess of enzyme. Southern blotting analysis was performed as previously described utilizing an apoA-I genomic probe (2.2 kb PstI fragment) radiolabeled with ^{32}P by the random primer method.

Statistical analysis

Statistical differences between groups with regard to allele frequencies were assessed by chi-square analysis. The relationships between plasma lipid levels and the presence or absence of various alleles were assessed using the SAS statistical package as previously described [21]. Variables with a non parametric distribution were $\log_{10}$ transformed.

RESULTS

Table 1 displays the lipid levels in males and females in the control Spanish population. The mean age was 34.0 years old, with a range of 18 to 62 years. Total cholesterol, triglyceride and HDL cholesterol were 185.7; 65.2; and 55.4 mg/dl respectively.

Table II summarizes previously published data regarding the SstI RFLP on Caucasians populations. The frequency of this RFLP in the Spanish population is also presented. The examination of all published reports, including 2,223 normal Caucasians, indicate that the average frequency of the rare allele at the SstI site was 0.072 ± 0.078 (SD), ranging from 0.00 in some of the studies in London, to 0.40 in a small population sample in Italy. 1,170 Caucasians with either lipid disorders,

Table I

LIPID AND LIPOPROTEIN PARAMETERS IN SPANISH CONTROLS

	MALES (151)		FEMALES (78)	
	mg/dl $\pm$ SD	(range)	mg/dl $\pm$ SD	(range)
AGE	35.1 $\pm$ 11.7	(18-62)	32.2 $\pm$ 9.1	(20-58)
TOTAL CHOL.	182.2 + 28.8	(124-297)	192.4 $\pm$ 30.1	(136-282)
HDL CHOLESTEROL	53.9 $\pm$ 8.3	(35-84)	58.4 $\pm$ 14.7	(30-145)
TRIGLYCERIDES	66.4 $\pm$ 33.9	(19-194)	62.9 $\pm$ 33.1	(18-213)

diabetes or CHD, were also studied. In this population the frequency of the rare allele (S2) was 0.14 ± 0.10 (SD), with a range of 0.00 to 0.50. The difference in allele frequency between cases and controls in Caucasians was highly significant ($p < 0.001$).

The results from previous studies in non-Caucasian populations are summarized in Table III. A total of 1,050 control subjects are presented. The ethnic groups studied include: Black, Coloured, Chinese, Japanese, Indians, Philipinos, Arabs, and American Indians. The average frequency of the rare SstI allele was 0.218 ± 0.127 (SD). The range was from a minimum of 0.02 in a small sample of Kuwait, to a maximum of 0.47 in a small population of Chinese. Two hundred and sixty nine subjects affected with different dyslipidemias and/or CHD are also presented. The allele S2 allele frequency in this case population was 0.268 ± 0.145. There was not significance difference between the allele frequencies in cases and controls ($p = 0.067$). The difference in allele frequency between Caucasian and non-Caucasian control populations was highly significant ($p < 0.001$).

Table IV displays the PstI data from this and previous studies in Caucasians and non-Caucasians populations. The results obtained by pooling a sample of 2,405 Caucasian controls indicate that the frequency of the rare allele at this site is 0.068 ± 0.032 (SD). The frequencies of the P2 allele range from 0.01 to 0.14. In 1,157 Caucasian cases with different forms of lipoprotein disorders and/or CAD the average frequency of the P2 allele was 0.096 ± 0.092 (SD), with a range of 0.00 to 0.37. This difference was statistically significant at the level of $p < 0.05$. With regard to non-Caucasian populations, a total of 408 controls and 30 cases were studied. Because the low number of cases compared to controls no assessment of significance was performed. In the populations studied, there was not any appreciable ethnic effect on the frequency of this RFLP.

DISCUSSION

A number of candidates genes that may play a role in the genetic regulation of lipoprotein metabolism have been isolated and characterized [55]. Genetic variability at these loci have been demonstrated by the use of RFLP analysis. A number of these RFLPs have been reported to be associated with lipoprotein abnormalities or with CHD risk. Two of these RFLPs at the apoAI-CIII-AIV gene locus (PstI and SstI) have been studied extensively following a number of promising initial reports. The allele defined by the addition of a SstI cutting site in the 3' untranslated region of the apoC-III gene (S2) has been associated with increased plasma triglycerides in some Caucasian populations. The overall frequency of the S2 allele in all Caucasians controls was 0.072. Many of these studies were performed in subjects from the London area. This population appears to have an extremely low frequency of the S2 allele in normolipemic individuals, and it could result in underestimation of the overall allele frequency in all Caucasian populations. Our own studies in Spanish and USA populations have resulted in an S2 allele frequency of 0.08. Another problem found in the assessment of the S2 allele frequency based on pooled data is the possibility that some of the subjects where reported repeatedly in different studies. Despite these objections, we estimate that

Table II

REFERENCE	GROUPS STUDIED		RESULTS			
	NORMAL	DISEASE	NORMAL	DISEASE	S2 ALLELE	D
			S1S1:S1S2:S2S2		NORM/DIS	
London [14]	Normo	IV/V	36:0:0	16:10:2	0.00/0.25	+
		IIb		7:0:0	0.00/0.00	0
London [22]		HTG/D		30:15:0	/0.17	
London [23]	Normo	HTG	52:0:0	48:23:3	0.00/0.19	+
London [24]	Random	MI	45:2:0	38:9:1	0.02/0.11	+
London [25]	Normo	II	64:9:0	22:6:0	0.06/0.11	0
		III		8:0:0	0.06/0.00	0
		IV		5:0:0	0.06/0.00	0
		V		6:4:1	0.06/0.27	+
Scotland [26]	Random	MI	89:28:0	54:11:0	0.12/0.08	0
London [27]	No CAD	Mod.CAD	64:3:1	34:4:0	0.04/0.05	0
		Sev.CAD		48:12:1	0.04/0.11	+
London [28]	Normo	III	75:6:0	17:2:0	0.04/0.05	0
Italy [29]	Random		6:12:2		0.40/	
London [30]	Random	MI	71:3:0	44:9:2	0.02/0.12	+
London [31]	Normo	HTG	30:1:0	26:12:0	0.02/0.16	+
		HC		47:11:0	0.02/0.10	+
Seattle [32]	Random	CHD	(101)	(140)	0.06/0.12	+
Iowa [33]	Random	CAD	30:6:0	34:7:2	0.08/0.13	+
London [34]	Normo		64:9:0		0.06/	
Norway [34]	Normo	HTG	22:11:0	20:6:0	0.12/0.17	0
Finnish [35]	Normo	CHD	51:10:0	30:9:0	0.08/0.11	0
		HTG		16:10:0	0.08/0.19	+
Vancouver [18]	Normo	FCH	33:4:1	18:5:1	0.05/0.15	+
		FH		14:1:0	0.05/0.03	0
		FD		10:4:0	0.05/0.12	0
London [36]	Normo	HTG	50:0:0	33:23:1	0.00/0.22	+
London [37]	Normo		91:1:0		0.01/	
S.Africa NAS [38]	Normo	HC	(37)	(21)	0.01/0.20	+
		HTG		(6)	0.01/0.17	0
		III		(1)	0.01/.50	0
S.Africa AS [38]	Normo	HC	(38)	(11)	0.12/0.14	0
London [39]	Random	CHD/D	34:1:0	34:13:0	0.02/0.14	+
Seattle [40]	Random		300:64:2		0.09/	
London [41]	Normo	CAD	48:2:0	37:12:0	0.02/0.12	+
Austria [42]	Random	CHD	93:23:2	84:22:0	0.11/0.10	0
Bristol [43]	Random		75:15:0		0.08/	
Greece [44]	Random		(129)		0.09/	
Boston [45]	Random		(160)		0.06/	
S.Arab [46]	Random		63:6:0		0.04/	
Boston [15]	Random	CAD	122:23:0	153:44:3	0.08/0.12	0
Spain	Random		149:25:2	22:12:1	0.08/0.20	+

Comments: Normal populations have been classified as normolipemics (normo) when subjects at the extremes of the distribution were excluded. Random indicates that no criteria for exclusion was indicated by the authors. For abbreviations see Table III.

Table III

FREQUENCY OF THE SstI RFLP IN PREVIOUS AND PRESENT STUDIES
(NON-CAUCASIANS)

REFERENCE	GROUPS STUDIED		RESULTS			
	NORMAL	DISEASE	NORMAL	DISEASE	S2 ALLELE	D
			S1S1:S1S2:S2S2		NORM/DIS	
Chinese [37]	Normo		7:7:6		0.47/	
Japanese [37]	Normo		13:8:0		0.19/	
Indians As. [37]	Normo		18:10:0		0.18/	
Africans [37]	Normo		14:6:0		0.15/	
Blacks [38]	Normo	HTG	37:5:0	1:0:0	0.06/0.00	
		III		4:1:0	0.06/0.13	
Coloured [38]	Normo	HC	34:9:0	20:15:1	0.10/0.24	+
		HTG		7:6:1	0.10/0.29	+
		III		3:2:1	0.10/0.33	0
Negroes [37]	Normo		15:11:2		0.27/	
Indians As. [37]	Normo		16:6:2		0.19/	
Japanese [37]	Normo		10:21:3		0.35/	
Japanese [47]	Normo	IIa	26:39:9	12:8:1	0.38/0.24	0
		IIb		2:5:2	0.38/0.50	0
		IV,V		11:11:2	0.38/0.31	0
		CAD		12:10:4	0.38/0.35	0
Japanese [40]	Random		24:38:6		0.37/	
Blacks [40]	Random		40:12:1		0.13/	
Blacks [44]	Random		(67)		0.05/	
Arabs [48]	Normo	HTG	30:1:0	11:26:1	0.02/0.37	+
Chinese [49]	Random	NIDDM	(73)	(93)	0.39/0.30	0
NW Indians [50]	Random		(130)		0.25/	
Philipinos [46]	Random		52:25:7		0.23/	
Arabs [46]	Random		102:38:7		0.18/	
Arabs [51]	Random	CAD	43:20:0	22:10:2	0.16/0.21	+

Familial hyperlipidemias are shown as described by the authors:
Types IIa, IIb, III, IV, V, hypertriglyceridemia (HTG), hypercholesterolemia (HC), familial hypercholesterolemia (FH), familial combined hyperlipidemia (FCH), familial dyslipidemia (FD). MI indicates myocardial infarction survivors, and D indicates diabetes. NAS indicates non afrikaneer speaker; and AS stands for afrikaneer speaker. The last column (D) indicates whether the authors reported significant association (+); no association (0) or negative association (-).
For comments see Table II.

Table IV

FREQUENCY OF THE PstI RFLP IN PREVIOUS AND PRESENT STUDIES

REFERENCE	GROUPS STUDIED		RESULTS			
	NORMAL	DISEASE	NORMAL	DISEASE	P2 ALLELE	D
			P1P1:P1P2:P2P2		NORM/DIS	
London (C) [25]	Normo	IIa	53:15:2	8:7:1	0.14/0.28	0
		IIb		9:1:0	0.14/0.05	0
		III		4:4:0	0.14/0.25	0
		IV		4:0:0	0.14/0.00	0
		V		8:2:0	0.14/0.10	0
Italy (C) [29]	Random	HA	19:1:0	4:7:1	0.02/0.37	+
Iowa (C) [33]	Normo	CAD	35:1:0	36:7:0	0.08/0.11	+
Seattle (C) [32]	Random	CAD	(114)	(140)	0.07/0.10	0
London (C) [30]	Random	MI	50:7:0	49:3:0	0.06/0.03	0
Boston (C) [16]	Random	CAD	118:5:0	60:26:2	0.02/0.17	+
Vancouver (C) [18]	Normo	FCH	33:5:0	23:1:0	0.06/0.02	0
		FH		14:1:0	0.06/0.03	0
		FD		12:2:0	0.06/0.07	0
London (C) [37]	Normo	HTG	55:5:0	27:9:0	0.12/0.04	0
London (C) [37]	Normo		16:4:0		0.10/	
London (B) [37]	Normo		26:3:0		0.05/	
London (IA) [37]	Normo		17:5:0		0.11/	
Tokyo (J) [37]	Normo		28:2:0		0.07/	
Austria (C) [42]	Random	CHD	105:11:2	95:11:0	0.06/0.05	0
Greece (C) [44]	Random		(129)		0.07/	
Baltimore (B) [44]	Random		(67)		0.03/	
Bristol (C) [43]	Random		88:13:1		0.07/	
London (C) [52]	Random	PVD	131:26:3	143:23:2	0.10/0.08	0
London (C) [52]	Random	CAD	91:12:0	87:19:4	0.06/0.12	+
London (A) [52]		CAD		23:7:0	0.06/0.12	0
Austral. (C) [53]	Random	CAD	82:11:0	93:7:0	0.04/0.06	0
Scotland (C)[54]	Random		601:73:5		0.06/	
Canada (NWI) [50]	Random		(260)		0.01/	
Boston (C) [45]	Random		(152)		0.06/	
Boston (C) [15]	Random	CAD	123:19:0	183:19:0	0.07/0.05	0
Spain (C) Present Study	Random		163:26:0		0.07/	

For abbreviations and comments see Tables II and III. HA stands for familial hypoalphalipoproteinemia; and PVD for peripheral vascular disease.
The different ethnic groups are abbreviated as follows: A, Asians; B, Blacks; C, Caucasians; IA, Indians Asians; J, Japanese; NWI, Northwest Indians.

the allele frequency in Caucasians is close 0.08. However, some regional differences could be observed among these populations. Scottish, Norwegians, and Afrikaneers seem to have a higher frequency of the S2 allele in the normal population, than English, Irish, Spanish or Greeks. This could be due to the small number of subjects analyzed in some of the studies, or it could be due to real genetic differences.

The frequency of the S2 allele in all cases reported in the literature, including different lipoprotein disorders, as well as diabetes and coronary heart disease, was significantly higher (0.13) than the one found in control subjects. These data suggest that this RFLP could be associated in Caucasians with some other mutation affecting the function of some of the genes in that locus. It is quite possible that the expression of the phenotype associated with this allele is dependent on a number of environmental factors and or other genes, as it has been demonstrated for the CETP locus [56,57]. In non-Caucasian control subjects there is a significantly higher frequency of the S2 allele (0.218) as compared to Caucasians (0.07). This difference is not exclusive of any specific ethnic group, and it can be demonstrated in Asians, Africans, and American Indians. In most of these populations, the S2 allele was not associated with either lipid disorders or increased CAD risk. It has been shown that some of ethnic groups are more affected by environmental changes than others. Recent studies have demonstrated that Asians immigrants have a higher risk of CAD than white populations in Britain [58]. Similar findings have also been reported in other populations [59]. Whether genetic variability at this locus, could be responsible for some of the differences in response to a new environment remains to be elucidated.

The rare allele of the PstI RFLP located 3' to the apoA-I gene was found in earlier reports to be strongly associated with increased risk of premature CAD and familial hypoalphalipoproteinemia. Other reports have demonstrated weaker associations in the same direction. However, most studies, including one from our group, have failed to demonstrate a significant association between the rare allele at this site and premature CAD. No differences were observed in allele frequencies at this site among the different ethnic groups studied. It is possible that the association between this RFLP and premature CAD it is exclusive to one of the several haplotypes that have been found to have this PstI site absent [44]. However, a more realistic scenario is that the positive associations previously found were obtained by chance, due to the small number of subjects analyzed and the low frequency of this RFLP in the population.

One of the common RFLPs found at the apoAI-CIII-AIV gene locus, the MspI site located at position -78 bp 5'to the apoA-I gene, has been found to affect gene regulation [60]. At the present time, none of the other RFLPs has been linked to regulatory or structural modification of any gene product. Based on the previously published data, there is enough evidence to indicate that genetic variability at this locus is associated with changes in plasma lipid levels.

Despite these associations, it is becoming increasingly clear that RFLPs at this locus are not useful clinical markers for CAD risk assessment in population studies. New techniques have been developed in the last few years that allow a more thorough and precise identification of DNA mutations and genetic variability [61,62]. These technical advances will contribute to the characterization of specific mutations affecting regulatory or structural domains of genes involved in the regulation of lipoprotein metabolism.

REFERENCES

1. Miller NE, Forde OH, Thelle DS. The Tromso Heart Study: high density lipoproteins and coronary heart disease: a prospective case control study. Lancet 1977; 2:767-772.

2. Miller GJ, Miller NF. Plasma high density lipoprotein concentration and development of ischemic heart disease. Lancet 1975; 1:16-20.

3. Miller NE. Associations of high-density lipoprotein subclasses and apolipoproteins with ischemic heart disease and coronary atherosclerosis. Am Heart J 1987; 113:589-597.

4. Castelli WP, Doyle JT, Gordon T. HDL cholesterol and other lipids in coronary heart disease: The Cooperative Lipoprotein Phenotyping Study. Circulation 1977; 55:767-772.

5. Brown MS, Goldstein JL. A receptor-mediated pathway for cholesterol homeostasis. Science 1986; 232:34-47.

6. Innerarity TL, Mahley RW, Weisgraber KH, Bersot TP, Krauss RM, Vega GL, Grundy SM, Friedl W, Davignon J, McCarthy BJ. Familial defective apolipoprotein B-100: A mutation of apolipoprotein B that causes hypercholesterolemia. J Lipid Res 1990; 31:1337-1349.

7. Karathanasis SK. Apolipoprotein multigene family: tandem organization of human apolipoprotein A-I, C-III and A-IV genes. Proc Natl Acad Sci USA 1985; 82:6374-6378.

8. Bruns GA, Karathanasis SK, Breslow JL. Human apolipoprotein AI-CIII gene complex is located in chromosome 11. Arteriosclerosis 1984; 4:97-104.

9. Schaefer EJ, Ordovas JM, Law S, Ghiselli G, Kashyap ML, Srivastava LS, Heaton WH, Albers JJ, Connor WE, Lemeshev Y, Segrest J, Brewer HBJr. Familial apolipoprotein A-I and C-III deficiency, variant II. J Lipid Res 1985; 26:1089-1101.

10. Norum RA, Lakier JB, Goldstein S, Angel A, Goldberg RB, Black WD, Noffze DK, Dolphin PJ, Edelglass J, Borograd DD, Alaupovic P. Familial deficiency of apolipoprotein A-I and C-III and precocious coronary artery disease. N Engl J Med 1982; 306:1513-1519.

11. Ordovas JM, Cassidy DK, Civeira F, Bisgaier CL, Schaefer EJ. Familial apolipoprotein A-I, C-III and A-IV deficiency and premature atherosclerosis due to deletion of a gene complex on chromosome 11. J Biol Chem 1989; 264:16339-16342.

12. Karathanasis SK, Ferris E, Haddad IA. DNA inversion within the apolipoproteins AI/CIII/AIV -encoding gene cluster of certain patients with premature atherosclerosis. Proc Natl Acad Sci USA 1987; 84:7198-7202.

13.   Karathanasis SK, Zannis VI, Breslow JL. Isolation and characterization of cDNA clones corresponding to two different human apoC-III alleles. J Lipid Res 1985; 26:451-456.

14.   Rees A, Stocks J, Shoulders CC, Galton DJ, Baralle FE. DNA polymorphism adjacent to human apoprotein A-I gene: Relation to hypertriglyceridemia. Lancet 1983; 1:444-446.

15.   Ordovas JM, Civeira F, Genest JJ, Craig S, Robbins AH, Meade T, Pocovi M, Frossard P, Masharani U, Wilson PWF, Salem D, Ward RH, Schaefer EJ. Restriction fragment length polymorphisms of the apolipoprotein A-I, C-III, A-IV gene locus: Relationships with lipids, apolipoproteins, and premature coronary artery disease. Atherosclerosis 1991;

16.   Ordovas JM, Schaefer EJ, Salem D, Ward RH, Glueck CJ, Vergani C, Wilson PWF, Karathanasis SK. Apolipoprotein A-I gene polymorphism associated with premature coronary artery disease and familial hypoalphalipoproteinemia. N Engl J Med 1986; 314:671-677.

17.   Wojclechowski AP, Farral M, Cullen P, Wilson TME, Bayliss JD, Farren B, Griffin BA, Caslake MJ, Packard CJ, Shepherd J, Thakker R, Scott J. Familial combined hyperlipidemia linked to the apolipoprotein AI-CIII-AIV gene cluster on chromosome 11q23-q24. Nature 1991; 349:161-164.

18.   Hayden MR, Kirk H, Clark C, Frohlich J, Rabkin S, McLeod R, Hewitt J. DNA polymorphism in and around the apoAI-CIII genes and genetic hyperlipidemias. Am J Hum Genet 1987; 40:421-430.

19.   Warnick R, Benderson J, Albers JJ. Dextran Sulfate-Mg precipitation procedure for quantitation of high density lipoprotein cholesterol. Clin Chem 1982; 28:1379-1388.

20.   Kan YW, Dozy AM. Antenatal diagnosis of sickle-cell anemia by DNA analysis of amniotic fluid cells. Lancet 1978; 2:910.

21.   Genest JJ, Ordovas JM, McNamara JR, Robbins AM, Meade T, Cohn SD, Salem D, Wilson PWF, Masharani U, Frossard P, Schaefer EJ. DNA polymorphisms of the apolipoprotein B gene in patients with premature coronary artery disease. Atherosclerosis 1990; 82:7-17.

22.   Jowett NI, Rees A, Williams LG, Stocks J, Vella MA, Hitman GA, Katz J, Galton DJ. Insulin and apolipoprotein A-I/C-III gene polymorphisms relating to hypertriglyceridemia and diabetes mellitus. Diabetologia 1984; 27:180-183.

23.   Rees A, Stocks J, Sharpe CR, Vella MA, Shoulders CC, Katz J, Jowett NI, Baralle FE, Galton DJ. Deoxyribonucleic acid polymorphism in the apolipoprotein AI-CIII gene cluster (Association with hypertriglyceridemia). J Clin Invest 1985; 76:1090-1095.

24. Ferns GAA, Ritchie C, Stocks J, Galton DJ. Genetic Polymorphisms of apolipoprotein C-III and Insulin in survivors of myocardial infarction. Lancet 1985; 2:300-303.

25. Kessling AM, Horsthemke B, Humphries SE. A study of DNA polymorphisms around the human apolipoprotein A-I gene in hyperlipidemic and normal individuals. Clin Genet 1985; 28:296-306.

26. Morris SW, Price WH. DNA sequence polymorphisms in the apoA-I/C-III gene cluster. Lancet 1985; 2:1127.

27. Rees A, Stocks J, Williams LG, Caplin JL, Jowett NI, Camm AJ, Galton DJ. DNA polymorphisms in the apolipoprotein C-III and insulin genes and atherosclerosis. Atherosclerosis 1985; 58:269-275.

28. Vella MA, Kessling AM, Jowett NI, Rees A, Stocks J, Wallis S, Galton DJ. DNA polymorphisms flanking the apoA-I and insulin genes and type III hyperlipidemia. Hum Genet 1985; 69:275-276.

29. Sidoli A, Giudici G, Soria M, Vergani C. Restriction fragment length polymorphisms in the AI-CIII gene complex occuring in a family with hypoalphalipoproteinemia. Atherosclerosis 1986; 62:81-87.

30. Ferns GAA, Galton DJ. Haplotypes of the human apoprotein AI-CIII-AIV gene cluster in coronary atherosclerosis. Hum Genet 1986; 73:245-249.

31. Shoulders CC, Ball MJ, Mann JI, Baralle FE. Genetic marker in apolipoprotein AI/CIII gene complex associated with hypercholesterolemia. Lancet 1986; 2:1286.

32. Deeb S, Failor A, Brown BG, Brunzell JD, Albers JJ, Motulski AG. Molecular genetics of apolipoproteins and coronary heart disease. Cold Spring Harbor Symp Quant Biol 1986; 51:403-408.

33. Anderson RA, Benda TJ, Wallace RB, Eliason SL, Lee J, Burns TL. Prevalence and association of apolipoprotein A-I linked DNA polymorphism: Results from a population study. Genet Epidemiol 1986; 3:385-397.

34. Kessling AM, Berg K, Mockleby E, Humphries SE. DNA polymorphisms around the apo A-I gene in normal and hyperlipidemic individuals selected for a twin study. Clin Genet 1986; 29:485-490.

35. Aalto-Setala K, Kontula K, Sane T, Nieminen M, Nikkila E. DNA polymorphisms of apolipoprotein AI/CIII and insulin genes in familial hypertriglyceridemia and coronary artery disease. Atherosclerosis 1987; 66:145-152.

36. Stocks J, Paul H, Galton DJ. Haplotypes identified by DNA RFLP in the AI-CIII-AIV gene region and hypertriglyceridemia. Am J Hum Genet 1987; 41:106-118.

37.   Paul H, Galton DJ, Stocks J. DNA polymorphic patterns and haplotype arrangements of the apoA-I, apoC-III, apo A-IV gene cluster in different ethnic groups. Hum Genet 1987; 75:264-268.

38.   Henderson HE, Landon SV, Michie J, Berger GM. Association of a DNA polymorphism in the apolipoprotein C-III gene with diverse hyperlipidemic phenotypes. Hum Genet 1987; 75:62-65.

39.   Trembath RC, Thomas DJB, Hendra TJ, Yunkin JS, Galton DJ. Deoxyribonucleic acid polymorphism of the apoprotein AI-CIII-AIV gene cluster and coronary heart disease in non-insulin-dependent diabetes. British Med J 1987; 294:1577-1578.

40.   Thompson EA, Deeb S, Walker D, Motulski AG. The detection of linkage disequilibrium between closely linked markers: RFLPs at the AI-CIII apolipoprotein genes. Am J Hum Genet 1988; 42:113-124.

41.   O'Connor G, Stocks J, Lumley J, Galton DJ. A DNA polymorphism of the apolipoprotein C-III gene in extracoronary atherosclerosis. Clin Sci 1988; 74:289-292.

42.   Paulweber B, Friedl W, Krempler F, Humphries SE, Sandhofer F. Genetic variation in the apolipoprotein AI-CIII-AIV gene cluster and coronary heart disease. Atherosclerosis 1988; 73:125-133.

43.   Kessling AM, Rajput-Williams J, Bainton D, Scott J, Miller NE, Baker I, Humphries SE. DNA polymorphisms of the apolipoprotein A-II and AI-CIII-AIV genes: A study in men selected for differences in High density lipoprotein cholesterol concentration. Am J Hum Genet 1988; 42:458-467.

44.   Antonarakis SE, Oettgen P, Chakravarti A, Halloran SL, Hudson RR, Feisee L, Karathanasis SK. DNA polymorphism haplotypes of the human apolipoprotein AI-CIII-AIV gene cluster. Hum Genet 1988; 80:265-273.

45.   Hegele RA, Hennekens CH, Breslow JL. Allele frequencies of Apolipoproteins A-I and A-II gene locus DNA polymorphisms in Boston-Based Whites. Hum Hered 1989; 39:174-178.

46.   Johansen K, Skotnicki A, Tan JCY, Kwaasi AA, Skotnicki M. Apolipoprotein A-I/C-III gene cluster polymorphism in Saudi Arabians, Filipinos, and Caucasians. Clin Genet 1990; 37:194-197.

47.   Abutarani H, Matsumoto A, Itoh H, Murase T, Takaku F, Itakura H. Deoxyribonucleic acid polymorphism in the apolipoprotein A-I gene: A study in a Japanese population. Jpn J Med 1988; 27:56-59.

48.   Tas S. Strong association of a single nucleotide substitution in the 3' untranslated region of the apolipoprotein C-III with common hypertriglyceridemia in Arabs. Clin Chem 1989; 35:256-259.

49.   Xiang K, Cox NJ, Sanz N, Huang P, Karam JH, Bell GI. Insulin-Receptor and apolipoprotein genes contribute to development of NIDDM in Chinese Americans. Diabetes 1989; 37:17-23.

50. Cole SA, Szathmary EJE, Ferrel RE. Gene and gene-product variation in the apolipoprotein A-I/C-III/A-IV cluster in the Dogrib Indians of the Northwest Territories. Am J Hum Genet 1989; 44:835-843.

51. Johansen K, Dunn B, Tan JCY, Kwaasi AA, Skotnicki A, Skotnicki M. Coronary artery disease and apolipoprotein A-I/C-III gene polymorphism: a study of Saudi Arabins. Clin Genet 1991; 39:1-5.

52. Wile DB, Barbir M, Gallagher J, Myant NB, Ritchie C, Thompson GR, Humphries SE. Apolipoprotein A-I gene polymorphisms: frequency in patients with coronary artery disease and healthy controls and association with serum apo A-I and HDL-cholesterol concentration. Atherosclerosis 1989; 78:9-18.

53. Dorow DS, Burke J, Goding JW. Assessment of a PstI polymorphism of the apolipoprotein A-I gene in Australian patients with coronary artery disease. Aust NZ J Med 1989; 19:677-681.

54. Price WH, Morris SW, Kitchin AH, Wenham PR, Burgon R, Donald PM. DNA restriction fragment length polymorphisms as markers of familial coronary heart disease. Lancet 1989; 1:1407-1411.

55. Lusis AJ. Genetic factors affecting blood lipoproteins: the candidate gene approach. J Lipid Res 1988; 29:397-429.

56. Kaprio J, Ferrell RE, Kottke BA, Sing CF. Smoking and reverse cholesterol transport: evidence for gene-environment interaction. Clin Genet 1989; 36:266-268.

57. Berg K, Kondo I, Drayna DT, Lawn RM. "Variability gene" effect of cholesteryl ester transfer protein (CETP) genes. Clin Genet 1989; 35:437-445.

58. Hughes LO, Raval U, Raftery EB. First myocardial infarctions in Asian and white men. British Med J 1989; 298:1345-1350.

59. Miller GJ, Kotecha S, Wilkinson WH, Wilkes H, Stirling Y, Sanders TA, Broadhurst A, Allison J, Meade TW. Dietary and other characteristics relevant for coronary heart disease in men of Indian, West Indian and European descent in London. Atherosclerosis 1989; 70:63-72.

60. Pagani F, Sidoli A, Giudici GA, Barenghi L, Vergani C, Baralle, FE. Human apolipoprotein A-I gene promoter polymorphism: association with hyperalphalipoproteinemia. J. Lipid Res. 1990; 31:1371-1378.

61. Boerwinkle E, Xiong W, Fourest E, Chan L. Rapid typing of tandemly repeated hypervariable loci by the polymerase chain reaction: Application to the apolipoprotein B 3' hypervariable region. Proc Natl Acad Sci USA 1989; 86:212-216.

62. Orita M, Suzuki Y, Sekiya T, Hayashi K. Rapid and sensitive detection of point mutations and DNA polymorphisms using the polymerase chain reaction. Genomics 1989; 5:874-879.

THE MOLECULAR BASIS OF THE CHYLOMICRONEMIA SYNDROME

Detlev Ameis[1], Junji Kobayashi[2], Heiner Greten[1], and Michael C. Schotz[2]

[1] Medizinische Kernklinik und Poliklinik, Universitäts-Krankenhaus Eppendorf Hamburg, F.R.G.
[2] Research VA Wadsworth Medical Center and Department of Medicine University of California, Los Angeles, CA 90073, U.S.A.

INTRODUCTION

Familial chylomicronemia (type I hyperlipoproteinemia) is a rare disorder of lipid metabolism characterized by a massive increase in plasma chylomicrons of fasting subjects resulting in a pronounced increase in plasma triglycerides [1]. Familial deficiencies of lipoprotein lipase (LPL) [2] and of apolipoprotein C-II [3], a polypeptide cofactor of LPL, are well-defined causes of this syndrome. LPL, the enzyme most frequently reduced or absent in chylomicronemia, is vital for the metabolism and transformation of triglyceride-rich lipoproteins circulating in the plasma [4,5]. Recently, both the complementary DNA [6,7] and the genomic structure [8-10] of LPL have been established. Based on this information, detailed studies of individuals with familial LPL deficiency are being carried out, utilizing gene amplification and nucleotide sequencing. Alterations in the LPL primary structure in some of these patients have been defined and shown to be the cause of the chylomicronemia syndrome.

LIPOPROTEIN LIPASE

LPL, the hydrolytic enzyme involved in a majority of cases with familial chylomicronemia, has a central role in the metabolism of triglyceride-rich lipoproteins. This enzyme is synthesized by the parenchymal cells of many tissues, most notably heart, muscle and adipose tissue, and is transported by an as yet unidentified mechanism to the luminal surface of vascular endothelial cells. In this location, it presumably binds to heparan sulfate [11,12], and hydrolyzes triglycerides of circulating chylomicrons and very-low-density lipoproteins, liberating free fatty acids for uptake by the parenchymal cells and further intracellular metabolism [4]. Intravenous injection of heparin at a dose of 60 units per kg of body weight releases LPL into the circulation where its enzymatic activity [13,14] and mass [15,16] can be determined. The absence or severe reduction of LPL activity in post-heparin plasma in the presence of apolipoprotein C-II establishes the diagnosis of LPL deficiency [1].

LPL has been successfully isolated from a number of mammalian sources [4,11]. The functional human LPL is a homodimeric protein with a protein subunit size of 448 amino acids [6]. A panel of human-hamster somatic cell hybrids was utilized to map the LPL gene to the short arm of chromosome 8 [17] and the recent isolation and characterization of genomic clones has shown it to contain 10 exons spanning about 30 kb of genomic DNA (Figure 1) [8,9]. Putative functional domains have been localized on the linear cDNA sequence (Figure 1), based on the recently elucidated three-dimensional structure of a related enzyme, human pancreatic lipase [18]. Comparisons of genomic and cDNA sequences indicate a remarkably high degree of homology to human hepatic lipase [19,20], pancreatic lipase [21] as well as to other lipases [22-26] and serine proteases[27]. The availability of human cDNA and genomic clones and genomic sequence information has made it possible to directly assess the molecular basis of LPL deficiency.

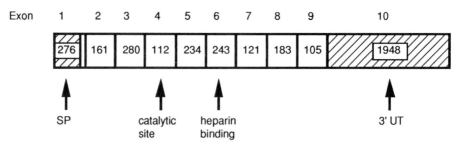

Fig. 1. Organization of the human LPL gene. Exon numbers are shown in the top row. The boxed region depicts the human LPL cDNA, with the vertical lines indicating exon-intron boundaries and the numbers indicating exon sizes in base pairs. The hatched areas denote the 5' and 3' untranslated cDNA sequences. SP, signal peptide; 3' UT, 3' untranslated sequence.

FAMILIAL LPL DEFICIENCY

Familial LPL deficiency is a rare autosomal recessive disorder of lipid metabolism characterized by a striking increase in plasma chylomicrons of fasting subjects and a concomittant increase in plasma triglyceride levels frequently above 17 mmol/l [1]. Clinically, this chylomicronemia is associated with a variety of symptoms, including recurrent abdominal pain, pancreatitis, hepatosplenomegaly, eruptive cutaneous xanthomas, and lipemia retinalis. In most instances the disease is detected in childhood. Restriction of dietary fat intake to 20 g or less is therapeutically used to control triglyceride levels and related clinical symptoms.

Deficiencies of LPL [2] and apolipoprotein C-II [3] have been well established as causes of the chylomicronemia syndrome, and both diseases have been extensively characterized biochemically [4,5,28,29]. Pronounced reduction or absence of LPL activity in post-heparin plasma or in adipose tissue biopsies confirm the diagnosis of LPL deficiency.

The molecular cloning of the cDNA for human LPL [6,7] and availablity of information on the genomic structure of the human LPL gene [8,9] makes it possible to directly investigate the molecular pathology of familial LPL deficiency, utilizing two different approaches. First, segregation of LPL mutations in families can be traced using restriction fragment length polymorphisms. A recent study demonstrated that gross alterations in the LPL gene are a major cause of primary LPL deficiency in a significant number of chylomicronemic subjects [30,31]. An alternative and more sensitive approach involves the direct visualization of deficiency-causing mutations in the LPL gene, employing polymerase chain reaction to specifically amplify genomic LPL sequences, and determination of the LPL DNA sequence. To exploit the latter approach, we studied a kindred with classical LPL deficiency as defined by severely reduced post-heparin plasma LPL activity and marked hypertriglyceridemia. Genomic amplification of LPL exons and direct DNA sequence determination yielded the genomic LPL sequence in two chylomicronemic siblings from a family of Northern European ancestry. A single base mutation at nucleotide position 680 on the human LPL cDNA was detected, resulting in replacement of glycine (codon GGA) by a glutamic acid residue (codon GAA) at amino acid position 142 of the mature LPL protein[32]. To trace the inheritance of the observed mutation through the family, three key members, the father, the mother, and the paternal grandfather were also analyzed. DNA sequence of these three family members showed heterozygosity for the observed mutation. The functional consequences of the mutation were confirmed using site-directed mutagenesis to introduce the codon-142 mutation in the normal LPL cDNA. Tissue culture expression studies utilizing a monkey kidney cell line (COS-7) transfected with the mutant LPL cDNA demonstrated synthesis of relatively normal amounts of LPL mass. However, the mutated LPL was not catalytically active, nor was it quantitatively secreted into the culture media. These results strongly suggest that the glutamic acid for glycine substitution at residue 142 has a major adverse effect on the tertiary structure of LPL resulting in a non-functional protein. An overview of the currently described LPL gene mutations is shown in Figure 2.

The recent molecular analysis of additional pedigrees has shown point mutations in other regions of the LPL molecule. Sequence determination of a kindred from Bethesda has revealed a single amino acid substitution of alanine for threonine at position 176, resulting in an inactive protein with reduced affinity for heparin [33]. A large family of Northern European descent was demonstrated to be LPL-deficient due to an amino acid substitution of glycine at position 188 to glutamic acid, also leading to a catalytically inactive LPL protein with reduced binding to heparin [34]. Investigations of further pedigrees have shown mutations at amino acids 106 [35], resulting in a truncated protein, and a heterozygous amino acid exchange at position 244 [36] responsible for LPL enzymatic deficiency.

The LPL gene mutations detected at different locations along the molecule suggest that LPL is very sensitive to single amino acid exchanges, resulting in loss of hydrolytic activity. Moreover, these mutations seem to be clustered on exons 3 through 6 of the LPL gene, indicating a particularly high constraint to sequence alterations in these regions. Identification of additional LPL mutants in different chylomicronemic subjects and the exploitation of site-directed mutagenesis to specifically modify the LPL protein will allow the detailed analysis of structure-function relationships and further broaden our understanding of their role in triglyceride metabolism.

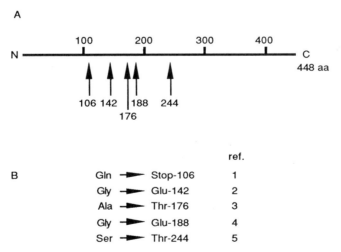

Fig. 2. Mutations of the LPL protein responsible for LPL deficiency. Panel A shows the location of amino acid exchanges on the LPL protein of five different individuals with the chylomicronemia syndrome. Panel B gives the corresponding amino acid exchanges. 1, ref. 35; 2, ref. 32; 3, ref. 33; 4, ref. 34; 5, ref. 36. N, amino terminal end of the mature LPL protein; C, carboxy terminal end.

Acknowledgement – These studies are being supported by grants from the National Institutes of Health (HL 28481) and the Veterans Administration. D.A. receives support by the Deutsche Forschungsgemeinschaft (Am 65/3-1) and J.K. received support by Mochida funds (Japan). We are grateful to Drs Mary Malloy, John Kane and Richard Havel for their many contributions and help.

REFERENCES

1. J.D. Brunzell, Familial lipoprotein lipase deficiency and other causes of the chylomicronemia syndrome, in: The metabolic basis of inherited disease, C.R. Scriver, A.L. Beaudet, W.S. Sly, and D. Valle, ed., McGraw-Hill, New York (1989) p. 1165.

2. R.J. Havel and R.S. Gordon,Jr., Idiopathic hyperlipemia: metabolic studies in an affected family, J Clin Invest 39:1777 (1960).

3. W.C. Breckenridge, J.A. Little, G. Steiner, A. Chow, and M. Poapst, Hypertriglyceridemia associated with deficiency of apolipoprotein C-II, N Engl J Med 298:1265 (1978).

4. A.S. Garfinkel and M.C. Schotz, Lipoprotein lipase, in: Plasma lipoproteins, A.M. Gotto,Jr., ed., Elsevier, New York (1987) p. 335.

5. R.H. Eckel, Lipoprotein lipase: a multifunctional enzyme relevant to common metabolic diseases, N Engl J Med 320:1060 (1989).

6. K.L. Wion, T.G. Kirchgessner, A.J. Lusis, M.C. Schotz, and R.M. Lawn, Human lipoprotein lipase complementary DNA sequence, Science 235:1638 (1987).

7. T. Gotoda, M. Senda, T. Gamou, Y. Furuichi, and K. Oka, Nucleotide sequence of human cDNA coding for a lipoprotein lipase (LPL) cloned from placental cDNA library, Nucl Acids Res 17:2351 (1989).

8. S.S. Deeb and R. Peng, Structure of the human lipoprotein lipase gene, Biochemistry 28: 4131 (1989).

9. T.G. Kirchgessner, J.-C. Chuat, C. Heinzmann, J. Etienne, S. Guilhot, K. Svenson, D. Ameis, C. Pilon, L. d'Auriol, A. Andalibi, M.C. Schotz, F. Galibert, and A.J. Lusis, Organization of the human lipoprotein lipase gene and evolution of the lipase gene family, Proc Natl Acad Sci USA 86:9647 (1989).

10. K. Oka, G.T. Tkalcevic, T. Nakano, H. Tucker, K. Ishimura-Oka, and W.V. Brown, Structure and polymorphic map of human lipoprotein lipase gene, Biochim Biophys Acta 1049:21 (1990).

11. T. Olivecrona and G. Bengtsson-Olivecrona, Lipoprotein lipase from milk - the model enzyme in lipoprotein lipase research, in: Lipoprotein lipase, J. Borensztajn, ed., Evener, Chicago (1987) p. 15.

12. C.-F. Cheng, G.M. Oosta, A. Bensadoun, and R.D. Rosenberg, Binding of lipoprotein lipase to endothelial cells in culture, J Biol Chem 256:12893 (1981).

13. P. Nilsson-Ehle and M.C. Schotz, A stable, radioactive substrate emulsion for assay of lipoprotein lipase, J Lipid Res 17:536 (1976).

14. H. Greten, R. Degrella, G. Klose, W. Rascher, J.L. de Gennes, and E. Gjone, Measurement of two plasma triglyceride lipases by an immunochemical method: studies in patients with hypertriglyceridemia, J Lipid Res 17:203 (1976).

15. J.W.F. Goers, M.E. Pedersen, P.A. Kern, J. Ong, and M.C. Schotz, An enzyme-linked immunoassay for lipoprotein lipase, Anal Biochem 166:27 (1987).

16. S.P. Babirak, P.-H. Iverius, W.Y. Fujimoto, and J.D. Brunzell, The detection and characterization of the heterozygote state for lipoprotein lipase deficiency, Arteriosclerosis 9:326 (1989).

17. R.S. Sparkes, S. Zollman, I. Klisak, T.G. Kirchgessner, M.C. Komaromy, T. Mohandas, M.C. Schotz, and A.J. Lusis, Human genes involved in lipolysis of plasma lipoproteins: Mapping of loci for lipoprotein lipase to 8p22 and hepatic lipase to 15q21, Genomics 1:138 (1987).

18. F.K. Winkler, A. D'Arcy, and W. Hunziker, Structure of human pancreatic lipase, Nature 343:771 (1990).

19. S.-J. Cai, D.M. Wong, S.-H. Chen, and L. Chan, Structure of the human hepatic triglyceride lipase gene, Biochemistry 28:8966 (1989).

20. D. Ameis, G. Stahnke, J. Kobayashi, M. Bücher, G. Lee, J. McLean, M.C. Schotz, and H. Will, Isolation and characterization of the human hepatic lipase gene, J Biol Chem 265:6552 (1990).
21. F.S. Mickel, F. Weidenbach, B. Swarowsky, K.S. LaForge, and G.A. Scheele, Structure of the canine pancreatic lipase, J Biol Chem 264:12895 (1989).
22. M.C. Komaromy and M.C. Schotz, Cloning of rat hepatic lipase cDNA: Evidence for a lipase gene family, Proc Natl Acad Sci USA 84:1526 (1987).
23. M. Senda, K. Oka, W.V. Brown, P.K. Qasba, and Y. Furuichi, Molecular cloning and sequence of a cDNA coding for bovine lipoprotein lipase, Proc Natl Acad Sci USA 84:4369 (1987).
24. T.G. Kirchgessner, K.L. Svenson, A.J. Lusis, and M.C. Schotz, The sequence of cDNA encoding lipoprotein lipase: A member of a lipase gene family, J Biol Chem 262:8463 (1987).
25. S. Enerbäck, H. Semb, G. Bengtsson-Olivecrona, P. Carlsson, M-L. Hermansson, T. Olivecrona, and G. Bjursell, Molecular cloning and sequence analysis of cDNA encoding lipoprotein lipase of guinea pig, Gene 58:1 (1987).
26. D.A. Cooper, J.C. Stein, P.J. Strieleman, and A. Bensadoun, Avian adipose lipoprotein lipase: cDNA sequence and reciprocal regulation of mRNA levels in adipose and heart, Biochim Biophys Acta 1008(1):92 (1989).
27. S. Brenner, The molecular evolution of genes and proteins: a tale of two serines, Nature 334:528 (1988).
28. P. Nilsson-Ehle, A.S. Garfinkel, and M.C. Schotz, Lipolytic enzymes and plasma lipoprotein metabolism, Ann Rev Biochem 49:667 (1980).
29. P.A. Kern, R.A. Martin, J. Carty, I.J. Goldberg, and J.M. Ong, Identification of lipoprotein lipase immunoreactive protein in pre- and postheparin plasma from normal subjects and patients with type I hyperlipoproteinemia, J Lipid Res 31:17 (1990).
30. S. Langlois, S. Deeb, J.D. Brunzell, J.J. Kastelein, and M.R. Hayden, A major insertion accounts for a significant proportion of mutations underlying human lipoprotein lipase deficiency, Proc Natl Acad Sci USA 86:948 (1989).
31. R.H. Devlin, S. Deeb, J. Brunzell, and M.R. Hayden, Partial gene duplication involving exon-Alu interchange results in lipoprotein lipase deficiency, Am J Hum Genet 46:112 (1990).
32. D. Ameis, J. Kobayashi, R.C. Davis, O. Ben-Zeev, M.J. Malloy, J.P. Kane, H. Wong, R.J. Havel, and M.C. Schotz, Familial chylomicronemia (type I hyperlipoproteinemia) due to a single missence mutation in the lipoprotein lipase gene, J Clin Invest, in press.
33. O.U. Beg, M.S. Meng, S.I. Skarlatos, L. Previato, J.D. Brunzell, H.B. Brewer, jr., and S.S. Fojo, Lipoprotein lipase (Bethesda): A single amino acid substitution (Ala-176 to Thr) leads to abnormal heparin binding and loss of enzymic activity, Proc Natl Acad Sci USA 87:3474 (1990).
34. M. Emi, D.E. Wilson, P.-H. Iverius, L. Wu, A. Hata, R. Hegele, R.R. Williams, and J.-M. Lalouel, Missence mutation (Gly to Glu-188) of human lipoprotein lipase imparting functional deficiency, J Biol Chem 265:5910 (1990).
35. M. Emi, A. Hata, M. Robertson, P.-H. Iverius, R. Hegele, and J.-M. Lalouel, Lipoprotein lipase deficiency resulting from a nonsense mutation in exon 3 of the lipoprotein lipase gene, Am J Hum Genet 47:107 (1990).
36. A. Hata, M. Emi, G. Luc, A. Basdevant, P. Gambert, P.-H. Iverius, and J.-M. Lalouel, Compound heterozygote for lipoprotein lipase deficiency: Ser to Thr-244 and transition in 3' splice site of intron 2 (AG to AA) in the lipoprotein lipase gene, Am J Hum Genet 47:721 (1990).

LIPOPROTEIN LIPASE GENE VARIANTS IN SUBJECTS WITH HYPERTRIGLYCERIDAEMIA AND CORONARY ATHEROSCLEROSIS

J.A. Thorn and D.J. Galton

Medical Professorial Unit, St. Bartholomew's
Hospital, West Smithfield, London EC1A 7BE

INTRODUCTION

The aggregation of coronary artery disease within families suggests a strongly inherited component in the predisposition to this disorder[1]. Analysis of first degree relatives of 121 men and 96 women with CAD showed the increased risk of death from the disease was five and seven fold greater than in matched controls for males and females respectively[2]. Many subsequent studies have confirmed this trend[3,4]. Further evidence for the implication of a genetic predisposition to CAD comes from twin studies[5]. Berg found concordance rates for angina pectoris or myocardial infarction to be 0.65 in monozygotic twins and 0.25 in dizygotic twins. If twins with premature CAD appearing before age sixty were considered alone these figures were 0.83 and 0.22 respectively[6]. Similarly, although hypertriglyceridaemia is known to be influenced by environmental factors such as diet, excessive alcohol intake and obesity, it has also been demonstrated to have a strong genetic component. Segregational analysis in certain pedigrees reveals a bimodal triglyceride distribution consistent with autosomal dominant inheritance[7].

There are a number of rare monogenic disorders such as familial lipoprotein lipase (LPL) deficiency and LDL receptor deficiency that make a small contribution to the inherited nature of hypertriglyceridaemia and CAD respectively. The modes of inheritance of the more common forms of these diseases do not, however, follow simple Mendelian rules. Instead, co-inheritance of several different variant genes in combination with the presence of appropriate environmental factors may be required.

The use of RFLPs to detect DNA sequence variation at candidate genes implicated in the inheritance of such polygenic conditions is well established. Alleles identified in this way at the apolipoprotein gene loci have been extensively studied for associations with both hypertriglyceridaemia and CAD. Gene variants at the apolipoprotein AI-CIII-AIV and CI-CII-E clusters have shown some associations with both disorders [8,9,10,11]. Other likely candidates are the genes encoding the enzymes responsible for the catabolism of plasma lipoproteins, including the lipolytic enzyme lipoprotein lipase (LPL).

Lipoprotein lipase activity is a rate determinant in the clearance of the triglyceride-rich lipoproteins VLDL and chylomicrons from the circulation [12]. Correlations have been reported between post-heparin LPL activity and the degree of post prandial lipaemia[13] which relates to the rate of clearance of these lipoproteins[14,15]. Some of the products of lipolysis of these lipoproteins, especially surface components, are transferred to high density lipoprotein particles resulting in a net conversion of the HDL_3 subfraction to HDL_2[16]. This may account for the observed positive correlation between post heparin plasma lipoprotein lipase activity and plasma HDL2 concentration[17]. Increased plasma duration of partially catabolised, atherogenic VLDL and chylomicron remnants associated with reduced production of HDL_2 as a result of low LPL activity may, therefore, contribute to the development of premature CAD. Data from work addressing a possible association between low LPL activity and prevalence of CAD is, however, equivocal. Probably, this reflects the many physiological factors affecting LPL synthesis and activation which may confound any underlying differences detected by the assay.

The human gene encoding this lipase has recently been cloned and mapped to human chromosome 8p22[18].It codes for a mature protein of 448 amino acids with a molecular weight of 65,000 Daltons[19]. The cloned gene has been used as a hybridisation probe to detect genomic DNA sequence variation in the form of restriction site polymorphisms. Digestion of genomic DNA with the restriction enzymes Hind-III[20] or Pvu-II[21] reveal separate two allele polymorphisms when probed with an LPL cDNA clone.

The current studies were designed to examine whether a gene variant at the LPL locus as defined by these RFLPs may associate with a lipoprotein phenotype indicating reduced catabolism of plasma triglycerides and, hence, with premature coronary atherosclerosis.

SUBJECTS

All subjects studied were unrelated and had fasting plasma glucose levels less than 6.5 mmol/l. The total cholesterol, triglyceride and HDL-cholesterol levels were measured in plasma from the fasting subjects. These are summarised in table I. Venous blood was taken from all individuals for DNA analysis, the sole subsequent reason for exclusion from the study was the lack of suitable DNA for digestion.

Hypertriglyceridaemia in Caucasian Subjects

Caucasian hypertriglyceridaemic subjects were recruited over a six month period from the lipid clinic at St. Bartholomew's hospital. They were selected on the basis of two pretreatment fasting measurements of plasma triglyceride >2.0mmol/l together with cholesterol <7.2mmol/l. HDL-cholesterol measurements were not available at the time of study.

TABLE I. CLINICAL DETAILS OF SUBJECTS

GROUP	n (SEX)	AGE/Yrs	FASTING PLASMA LIPIDS (mmol/l)		
			CHOLESTEROL	TRIGLYCERIDES	HDL-CHOLESTEROL
London Caucasian Hypertriglyceridaemic	41M 5F	50±15	6.8±1.0	6.9±5.0	Not Available
London Caucasian Normotriglyceridaemic Controls	93M	48±12	5.6±0.9	1.1±0.4	Not Available
Japanese Hypertriglyceridaemic	27M 2F	50±9	6.1±1.3	7.6±14.7	Not Available
Japanese Normotriglyceridaemic Controls	36M 6F	45±11	5.9±1.1	1.2±0.5	Not Available
London Caucasian CAD	63M 7F	57±6	6.5±1.3	2.4±1.9	0.9±0.3
London Caucasian With No Clinical Signs of CAD	112M 10F	48±12	5.8±1.0	1.2±0.5	1.3±0.3
Welsh Caucasian Severe CAD	60M	54±7	6.4±1.2	2.4±1.4	0.8±0.2
Welsh Caucasian, Normal Coronary Arteries	60M	54±9	5.8±1.0	1.8±1.0	0.9±0.2
Welsh Caucasian, Normal Coronary Arteries, Triglycerides <2.0mM	40M	55±9	5.6±1.0	1.4±0.3	0.9±0.2

A group of healthy Caucasian males was recruited from a
health screening clinic at the BUPA medical centre, London.
Control subjects (n=93) were selected from these on the basis of
fasting plasma triglycerides <2.0mmol/l and fasting plasma
cholesterol <7.2mmol/l.

Hypertriglyceridaemia in Japanese Subjects

Japanese hypertriglyceridaemic subjects (n=42) were
recruited from the University Hospital, Bunkyu, Tokyo and were
selected on similar criteria as the Caucasians.

Japanese controls (n=29) were recruited from the same
institute and from the Nippon Clinic in London, selected for
fasting plasma triglyceride measurements <2.0 mmol/l and fasting
plasma cholesterol <7.2 mmol/l.

Coronary Atherosclerosis in London Caucasian Subjects

Caucasian subjects who had developed severe coronary
atherosclerosis before the age of 60 years (n=70) were recruited
from St. Bartholomew's Hospital department of cardiology. The
extent of coronary atherosclerosis was defined by invasive
radio-angiography using the techniques of Sones and Judkins[22].
Individuals with angiographic score[23] greater than 15/30 were
included in the study. Many (n=55) were post operative coronary
artery bypass patients with two or three vessel occlusions.

Healthy caucasian male controls (n=112) were selected from
those recruited from the health screening clinic at the BUPA
Centre London and Caucasian females (n=10) serially selected
from St. Bartholomew's hospital outpatient clinics. None had

clinical signs, family or past history of ischaemic heart disease. All BUPA controls were selected on the basis of normal resting twelve-lead ECGs.

Coronary Atherosclerosis in Welsh Subjects

Over a period of fifteen months, 1371 Welsh Caucasian subjects were recruited during angiographic investigation for reported chest pain at the Department of Cardiology, University of Wales College of Medicine, Cardiff. The extent of any coronary atherosclerosis was defined by invasive radio-angiography using the techniques of Sones and Judkin[22]. They were then classified as possessing normal coronary arteries, or coronary atherosclerosis to a moderate or severe extent by angiographic scoring[23] carried out by a senior member of staff using a blinded protocol. Individuals from the severe CAD (n=60) and normal coronary arteries (n=60) groups were selected for further study. In addition, a group of subjects with fasting plasma triglycerides <2.0 mmol/l were selected from the group with normal arteries (n=40). All subjects were male, older than forty years in the case of those with normal coronary arteries and less than sixty five years for those with severe CAD. None had received any hypolipidaemic therapy or dietary advice except for beta-blocker or thiazide treatment for symptoms of CAD.

The experimental design was such that the genotyping of all subjects took place at St. Bartholomew's Hospital. Only after assignation of genotype was clinical data made available and statistical analysis undertaken.

DNA Analysis

DNA was isolated from fresh or frozen whole blood as described in detail elsewhere[8]. DNA samples of 8ug each were digested separately with the restriction endonucleases Hind-III and Pvu-II according to the manufacturer's recommendations (Gibco U.K.). Spermidine was incorporated at a concentration of 1-2 mmol/l to enhance digestion. The digests were electrophoresed on 0.85% agarose gels and Southern blotted on to "Hybond-N" membranes (Amersham International, Amersham, Bucks). The filters were hybridised with 5 X 106 cpm/ml 32P labelled probe using hybridisation conditions as previously described[8]. Following hybridisation and washing, the bands were visualised by autoradiography at -70^{0}C with pre-flashed "Hyperfilm-MP" (Amersham International) and intensifying screens.

The lipoprotein lipase probe was a 1.6 Kb cDNA clone[24] corresponding to nucleotides 64-1638[19] cloned in a pIBI vector. The probe was cut from its vector with EcoR-1, and separated on a low melting point agarose gel. The probe was radiolabelled by the random priming procedure[25].

Lipid Analysis

Venous blood was withdrawn from the subjects after a twelve to sixteen hour fast and mixed with 1mg/ml disodium EDTA. Plasma was obtained by centrifugation at 2,000 rpm for 10 minutes. Triglycerides and cholesterol were measured by fully enzymatic methods (Boehringer Mannheim, FRG.). HDL-cholesterol was measured by the heparin/MnCl$_2$ precipitation method.

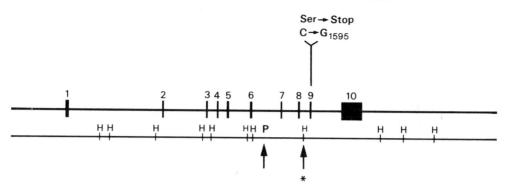

RESTRICTION MAP OF THE HUMAN LPL GENE

Figure 1. Map of the lipoprotein lipase gene showing
informative polymorphisms (P = Pvu II; H =
Hind III) and the exon 9 mutation producing
a premature stop codon.

Statistical Analysis

Genotype frequencies in both patient and control groups
were analysed by 2 X 3 contingency tables by chi squared analysis using the Minitab statistical package. Allele frequencies
and proportions of subjects with at least one allele were
compared by z-tests.

RESULTS

The map in figure 1 is of the human lipoprotein lipase gene
showing intron/exon structure and cutting sites of the restriction enzymes Hind-III and Pvu-II. The polymorphic sites are
marked by arrows. The Hind-III polymorphic site is in intron 8[26]
and the Pvu-II site is in intron 6, upstream of an ALU repeat
sequence[27].

Polymorphisms

Digestion with Hind-III reveals a two allele polymorphism
with an invariant band at 4.5Kb and a band at either 17.5Kb in
the absence of the restriction site (H1 allele) or 8.7Kb in its
presence (H2 allele)20. Pvu-II also reveals a two allele
polymorphism with an invariant band at 2.7Kb and a band at 7.0Kb
(P1 allele) which is cut to produce bands at 4.4Kb and 2.5Kb (P2
allele) when the polymorphic site is present [21].

LPL Genotypes in Caucasian and Japanese Hypertriglyceridaemic and Control Subjects

There is a strong association of the H2H2 genotype with
hypertriglyceridaemia, it being present in 65% of Caucasian

115

TABLE II. LPL GENOTYPE AND ALLELE FREQUENCIES IN HYPER-
TRIGLYCERIDAEMIA AND CONTROLS: LONDON CAUCASIAN AND JAPANESE
SUBJECTS

| GROUP | NUMBER OF SUBJECTS WITH EACH GENOTYPE (%) | | | | | | |
| | Hind-III | | | Pvu-II | | | |
	n	H1H1	H1H2	H2H2	n	P1P1	P1P2	P2P2
London Caucasian Hyper-triglyceridaemic	45	3(7)	13(28)	29(65)	46	11(24)	19(41)	16(34)
London Caucasian Normo-triglyceridaemic	93	17(18)	43(46)	33(36)	86	19(22)	41(48)	26(30)
Japanese Hyper-triglyceridaemic	29	2(7)	3(10)	24(83)	29	1(3)	4(14)	24(83)
Japanese Normo-triglyceridaemic	38	2(5)	22(58)	14(37)	41	3(7)	17(42)	21(51)

| GROUP | ALLELIC FREQUENCIES | | | | | |
| | Hind-III | | | Pvu-II | | |
	n	H1	H2	n	P1	P2
London Caucasian Hypertriglyceridaemic	90	0.211	0.789	92	0.446	0.554
London Caucasian Normotriglyceridaemic	186	0.414	0.586	172	0.459	0.541
Japanese Hypertriglyceridaemic	58	0.121	0.879	58	0.103	0.896
Japanese Normotriglyceridaemic	76	0.342	0.658	82	0.281	0.720

patients with the disorder compared to 36% of Caucasian controls
(table II) ($p<0.01$). No difference in the distribution of Pvu-II
genotypes was observed between the two groups. Similarly, the H2
allele frequency is raised in the Caucasian hypertriglyceridaemic group with respect to the controls ($p<0.01$).
The Pvu-II allelic frequencies are almost identical in the
patients and controls.

The association of the H2H2 genotype with hypertriglyceridaemia is also seen in the Japanese study groups. 83%
of the hypertriglyceridaemics possess this genotype compared
with 37% of the controls ($p<0.001$). In addition, the P2P2
genotype is also significantly more frequent in the Japanese
hypertriglyceridaemic group ($p<0.02$). The latter association may
reflect the greater linkage disequilibrium between the two
polymorphisms in this racial group ($D/D_{max}=95\%$) than in the
Caucasians ($D/D_{max}=23\%$). The H2 allele frequency is significantly
greater in the Japanese hypertriglyceridaemics than in the
controls ($p<0.01$), however the P2 frequency is not, despite the
genotype frequency difference. This may be due to the low number
of Japanese subjects that were available for study.

LPL Genotypes in Healthy Caucasian Subjects and Those With CAD

Table III presents the genotype and allele frequency data

TABLE III. LPL GENOTYPE AND ALLELE FREQUENCIES IN CAUCASIAN CAD AND HEALTHY SUBJECTS

| GROUP | NUMBER OF SUBJECTS WITH EACH GENOTYPE (%) | | | | | | | |
| | Hind-III | | | | Pvu-II | | | |
	n	H1H1	H1H2	H2H2	n	P1P1	P1P2	P2P2
London Caucasian CAD	63	3(5)	23(36)	37(59)	60	6(10)	39(65)	15(25)
London Caucasian, No Clinical signs of CAD	108	20(18)	51(47)	37(34)	93	20(21)	43(46)	30(32)

| GROUP | ALLELIC FREQUENCIES | | | | | |
| | Hind-III | | | Pvu-II | | |
	n	H1	H2	n	P1	P2
London Caucasian CAD	126	0.230	0.770	120	0.425	0.575
London Caucasian, No Clinical signs of CAD	216	0.421	0.579	186	0.446	0.554

for both polymorphisms in the London Caucasian CAD and control groups. On comparison of Pvu-II genotype frequencies in the two groups a significant difference is seen with a greater proportion of heterozygous individuals in the CAD group. There is, however, little difference in allelic frequencies between that in the control group (P2 0.554) and that in the CADs (P2 0.575). The Hind-III polymorphism shows a significant increase of the H2H2 genotype frequency in the CAD group. This is also seen on examination of allelic frequencies where there is a highly significant association of the H2 allele with the disease (H2 in controls 0.579, H2 in CADs 0.770; $p<0.001$).

Upon division of the CAD group into normolipidaemic and hyperlipidaemic subgroups, the H2 allele is significantly more common in individuals with total plasma cholesterol >6.5mmol/l (n=23) than in controls. This is also the case for individuals with plasma triglyceride >2.00 mmol/l (n=32) or plasma HDL-cholesterol <0.90mmol/l (n=27). The H2 allelic frequency in normocholestrolaemic or normotriglyceridaemic CAD patients does not differ significantly from that in the controls. Although the atherogenic risk conferred by hyperlipidaemia is lower for women than for men these observations were not influenced by exclusion of the women from the analysis. Within the CAD group there is a preponderance of mixed hyperlipidaemic individuals (31%) and the association seen between the H2 allele and raised plasma cholesterol is not independent from plasma triglyceride levels.

LPL Genotypes in Welsh Subjects With Normal Coronary Arteries or Severe CAD

The genotype and allele distributions of the Hind-III polymorphism in the Welsh Caucasian study groups are shown in table IV. Although there is a trend towards increased number of genotypes containing the H2 allele in the severe CAD as compared with the normal coronary arteries group, this does not quite reach significance (p<0.10 chi-squared). If, however, comparison is made between this group with severe CAD and those with normal coronary arteries selected for normotriglyceridaemia (fasting plasma triglycerides below 2.00 mmol/l) then a significant

TABLE IV. LPL GENOTYPE AND ALLELE FREQUENCIES AND PROPORTIONS OF INDIVIDUALS WITH AT LEAST ONE ALLELE IN SUBJECTS FROM WALES WITH SEVERE CAD OR NORMAL CORONARY ARTERIES

GROUP	NUMBER OF SUBJECTS WITH EACH GENOTYPE (%)				ALLELIC FREQUENCIES			PROPORTIONS OF SUBJECTS WITH AT LEAST ONE ALLELE		
	n	H1H1	H1H2	H2H2	n	H1	H2	n	H1	H2
Welsh Caucasian Severe CAD	60	1(2)	22(37)	37(61)	120	0.19	0.81	60	0.383	0.983
Welsh Caucasian Normal Coronary Arteries	60	7(12)	21(35)	32(53)	120	0.29	0.71	60	0.467	0.883
Welsh Caucasian Normal Coronary Arteries Triglycerides <2.0mM	40	7(17.5)	11(27.5)	22(55)	80	0.31	0.69	40	0.450	0.825

preponderance of the H2H2 genotype is demonstrated (p< 0.025 chi-squared). The frequency of the H2 allele was higher in this Welsh group with normal coronary arteries than observed in the previous control groups.

In a similar fashion the proportion of subjects with at least one H2 allele is significantly raised in the severe CAD as compared with the normal coronary arteries group (p< 0.05 Z-test). This trend is more pronounced if comparison is made between subjects with severe CAD and normotriglyceridaemic subjects with normal coronary arteries (p<0.005 Z-test).

DISCUSSION

The results from the series of studies described demonstrate a consistent association of the H2 allele of a Hind-III RFLP at the LPL gene locus with both primary hyper-triglyceridaemia and premature CAD. A criticism that is often directed against such population studies is that apparent differences in allele frequency between disease and control groups may reflect imperfect racial matching of the two rather than a genuine disease association. So if this work is to be taken seriously, then replication of an association in independent subject groups is essential. That the H2 allele was more frequent in four independent subject groups with hyper-triglyceridaemia and/or CAD than in matched controls, despite racial frequency differences between groups, suggests the association may reflect a physiological difference.

Some epidemiological studies have shown that hyper-triglyceridaemia is not an independent risk factor in the development of atherosclerosis when adjustments are made for HDL levels. The results of such multivariate analysis are, however, questionable as triglyceride and HDL may not be separate risk factors since their catabolic fates are closely linked. Recent data have shown the high triglyceride\low HDL syndrome to be the most common form of hyperlipidaemia amongst CAD patients[28]. Similarly the CAD groups in the present study contain a high proportion of hypertriglyceridaemics, many with a reduced HDL cholesterol. It is possible, therefore, that this hyperlipidaemia syndrome predisposes to the development of CAD. Thus a gene variant that influences clearance of plasma triglyceride and the metabolism of HDL may also affect the risk of premature coronary atherosclerosis. Our data now suggest that such a variant may be linked to the H2 allele at the lipoprotein lipase gene locus as it is associated with both primary hyper-triglyceridaemia and CAD. That the latter associations are attributable largely to the proportion of hypertriglyceridaemics in the CAD groups reinforces this concept.

Work is now underway to analyse LPL gene variants in pairs of siblings affected with CAD in order to determine whether haplotypes at this locus are shared more often than by chance. This will test the present association in the context of a pedigree based study, essential when investigating an inherited disease.

Individuals possessing this allele may have a gene variant that gives rise to altered lipoprotein lipase expression or activity. In view of the weakness of the association, the allelic frequencies in control groups and the intronic site of the polymorphism, it would seem unlikely that the H2 allele is in itself functional. It may, however, be a neutral marker in linkage disequilibrium with an aetiological locus in the lipoprotein lipase gene or an adjacent gene which predisposes to hypertriglyceridaemia and atherosclerosis.

A recently identified C-G transitional mutation in exon 9, adjacent to the Hind-III polymorphism, results in a premature stop codon, predicting a mature protein truncated by two amino acids[29]. The frequency of the allele that predicts the larger protein form was reported to be greater in a hyper-triglyceridaemic group (91%) than in normolipidaemic controls (67%). In view of this and the exonic location of the mutation, it is possible that this may represent a functional gene vari-ant. Further work is, therefore, required to ascertain whether the H2 allele is acting as a marker for this exon 9 mutation.

Phenotypic studies of LPL activity and secretion will then be necessary to confirm that this or any other subsequently identified common LPL protein polymorphism is functional. Such work may be able to define the role of LPL and the low HDL/high triglyceride syndrome in the aetiology of coronary atherosclerosis.

ACKNOWLEDGEMENTS

The authors gratefully acknowledge financial support from the Medical Research Council (UK) and the Joint Research Board of St. Bartholomew's Hospital.

REFERENCES

1.   Yater WM, Traum AH, Brown WG, Fitzgerald RP, Geisler MA,
Wilcox BB. Coronary artery disease in men eighteen to thirty-
nine years of age. Am Heart J 1948; 36, 334-72.
2.   Slack J, Evans KA. The increased risk of death from
ischaemic heart disease in first-degree relatives of 121 men and
96 women with ischaemic heart disease. J Med Genet 1966; 3, 239-
59.
3.   Rissanen AM. Familial aggregation of coronary heart disease
in a high incidence area (North Karelia, Finland). Br Heart J
1979; 42, 294-303.
4.   Nora JJ, Lortscher RH, Spangler RD, Nora AH, Kimberling WJ.
Genetic-epidemiologic study of early onset ischaemic heart
disease. Circulation 1980; 61, 503-8.
5.   Berg K. Twin studies of coronary heart disease and its risk
factors. Acta Genet Med Gemellol 1984; 33, 349-61.
6.   Berg K. Genetics of coronary heart disease. Prog Med Genet
(ed. by Steinberg AE, Bearn AG, Motulsky AR, Childs B). 1983;
35, 35-90.
7.   Goldstein JL, Schrott HG, Hazzard WR, Bierman EL, Motulsky
AG. Hyperlipidaemia in coronary heart disease II. Genetic
analysis of lipid levels in 176 families and delineation of a
new inherited disorder, combined hyperlipidaemia. J Clin Invest
1973; 52,
8.   Rees A, Shoulders CC, Stocks J, Galton DJ, Baralle FE. DNA
polymorphisms adjacent to human apolipoprotein A-I gene:
relation to hypertriglyceridaemia. Lancet 1983; i, 444-6.
9.   Rees, A., Stocks, J., Williams, L.G. et al. DNA Polymorphisms in the Apolipoprotein CIII and Insulin Genes and
Atherosclerosis. Atherosclerosis, 58 (1985) 269.
10.   Cumming, A.M. and Robertson, F.W. Polymorphism at the
Apoprotein-E Locus in Relation to Risk of Coronary Disease.
Clin. Genet., 25 (1984) 310.
11.   Lenzen, H.J., Assman, G., Buchwalsky, R. et al. Association of Apolipoprotein E Polymorphism, Low-Density Lipoprotein
Cholesterol and Coronary Artery Disease. Clin. Chem., 32 (1986)
778.
12.   Nilson-Ehle, P., Garfunkel, A.S. and Schotz, M.C. Lipolytic
Enzymes and Plasma Lipoprotein Metabolism. Ann. Rev. Biochem.,
49 (1980) 667.
13.   Nikkila E. Familial Lipoprotein lipase deficiency and
related disorders of chylomicron metabolism. The metabolic basis
of inherited disease. (ed. by Stanbury JB, Wyngaarden JB,
Fredrickson DS, Goldstein JL, Brown MS). Fifth edition 1983;
chapter 30. New York, McGraw-Hill.
14.   Taylor KG, Holdsworth G, Galton DJ. Lipoprotein lipase in
adipose tissue and plasma triglyceride clearance in patients
with primary hypertriglyceridaemia. Europ J Clin Invest 1980;
10, 133-38.
15.   Pykalisto OJ, Smith PH, Brunzell JD. Determinants of
adipose tissue Lipoprotein lipase. J Clin Invest 1975; 56, 1108-
16.   Patsch JR, Gotto AM Jr, Olivecrona T and Eisenberg S.
Formation of High Density Lipoprotein$_2$-like particles during
lipolysis of Very Low Density Lipoproteins in vitro. PNAS USA
1978; 75: 9: 4519-23.
17.   Patsch JR, Prasad S, Gotto AM,Jr, Patsch W. HDL$_2$ relationships of the plasma levels of this lipoprotein species to its
composition, to the magnitude of postprandial lipaemia and to

the activities of Lipoprotein lipase and Hepatic lipase. J Clin Invest 1987; 80, 341-7.

18. Sparkes RS, Zollman S, Klisak I et al. Human genes involved in the lipolysis of plasma lipoproteins: Mapping of the loci for Lipoprotein lipase to 8p22 and Hepatic lipase to 15q21. Genomics 1987; 1, 138-44.

19. Wion KL, Kirchgessner TG, Lusis AJ et al. Human Lipoprotein lipase complementary DNA sequence. Science 1987; 235, 1638-41.

20. Heinzmann, C., Ladias, J., Antonarakis, S. et al. RFLP for the Human Lipoprotein Lipase (LPL) Gene: Hind-III. Nucleic Acids Res., 15 (1987) 6763.

21. Fisher, L., Fitzgerald, G.A. and Lawn, R.M. Two Poly-morphisms in the Human Lipoprotein Lipase Gene. Nucleic Acids Res., 15 (1987) 7657.

22. Leamann DM, Browner RW, Meester GT et al. Coronary artery atherosclerosis: severity of the disease, severity of angina pectoris and compromised left ventricular function. Circulation 1981; 63, 285-92.

23. Brandt PWT, Partridge GB, Wattie WJ. Coronary arteriog-raphy; method of presentation of the arteriogram report and a scoring system. Clin Radiol 1977; 28, 361-5.

24. Gotoda T, Senda M, Gamou T et al. Nucleotide sequence of human cDNA coding for a Lipoprotein lipase (LPL) cloned from a placental cDNA library. Nucleic Acids Res 1989; 17, 6, 2351.

25. Feinberg, A.P. and Vogelstein, B. A Technique for Radio-labelling DNA Restriction Endonuclease Fragments to High Speci-fic Activity. Anal Biochem., 132 (1983) 6.

26. Stocks J, Galton DJ, Oka K and Oka KI. Human lipoprotien lipase HindIII and PvuII RFLPs revealed by the polymerase chain reaction. Nucleic Acids Res. 1990; Submitted.

27. Oka, K., Tkalcevic, G.T., Stocks, J. et al. Nucleotide Sequence of Pvu-II Polymorphic Site at the Human Lipoprotein Lipase Gene Locus. Nucleic Acids Res., 17 (1989) 6752.

28. Barbir, M., Wile, D., Trayner, I. et al. High Prevalence of Hypertriglyceridaemia and Apolipoprotein Abnormalities in Coronary Artery Disaese. Br. Heart J., 60 (1988) 397.

29. Hata M, Robertson M, Emi M and Lalouel J-M. Direct detec-tion and automated sequencing of individual alleles after electrophoretic strand separation: identification of a common nonsense mutation in exon 9 of the human LPL gene. Nucleic Acids Res. 1990; 18, 18, 5407-11.

ATHEROSCLEROSIS: THE GENETIC ANALYSIS OF A MULTI-FACTORIAL DISEASE

J.C. Chamberlain and D.J. Galton

Department of Human Metabolism and Genetics, St. Bartholomew's
Hospital, West Smithfield, London EC1A 7BE, UK

INTRODUCTION

Coronary atherosclerosis is a major cause of death in Western society and
its pathology is as complex as it is important. There is no one agent
responsible for all such atherogenesis, it is mostly a multifactorial disease
and it can be the end product of many influences, both environmental and
genetic.

The aggregation of premature coronary artery disease within families is
well known.[1] In 1966 Slack and Evans analysed first degree relatives of 121
men and 96 women with coronary artery disease (CAD) [2] and showed the increased
risk of death from CAD in these relatives. Since then many excellent studies
have confirmed this trend [3,4] and further evidence comes from twin studies
showing the concordance rates for angina pectoris or myocardial infarction in
monozygotic twins to be higher than for dizygotic twins (0.65 vs 0.25).[5] If
twins with premature CAD appearing before age 60 are alone considered these
differences are then even more marked (0.83 vs 0.22).[6]

THE USE OF LINKAGE MARKERS IN THE ANALYSIS OF POLYGENIC DISEASE

One of the major problems of understanding polygenic disorders such as
atherosclerosis is to distinguish between the inherited and secondary
components of the disease. Before the advent of recombinant DNA technology
there were no means to identify a mutant gene unless it produced a variant
protein. Now there is a possibility of directly studying the genes thought to
be involved in the aetiology of the disorder (the so-called candidate genes)
without recourse to a phenotypic intermediate.

The modes of inheritance of premature atherosclerosis are, if we exclude
the rarer monogenic disorders, complex and multifactorial, with inherited
"susceptibility" genes interacting with environmental factors to produce the
phenotypic disease. Only when genetic liability coincides with environmental

DNA Polymorphisms as Disease Markers, Edited by D.J. Galton and
G. Asmann, Plenum Press, New York, 1991

risk factors does the disease phenotype emerge. Realising this can help to explain the more puzzling features of the genetics of atherosclerosis. Its relatively high incidence may for example be due to the fact that under more favourable and natural environmental conditions the susceptibility genes are under no selective disadvantage and may instead offer other secondary advantages. In the face of such complexities the failure of classical genetics to elucidate recognisable patterns of inheritance is not surprising.

Our eventual goal must be the identification of these susceptibility loci for atherosclerosis and the quantification of their effect and environmental interaction, in the hope of predicting an individual's relative risk of developing the phenotypic disease.

Candidate Genes

In studying polygenic disease one question becomes paramount, "Which of the 1.4 million genes within the human genome are involved in the pathogenesis of the disease in question ?". Answers to this question can be achieved by one of two differing but complementary approaches.

1. Candidate gene targeting: This assumes that those genes known to code for a protein suspected as being involved with the disease pathology are the most worthy of study. In the case of atherosclerosis this approach might involve concentrating on a gene producing a protein known to be central to lipid metabolism ,on the assumption that hyperlipidaemia represents an intermediate phenotype for the development of atheroma, searching for mutation at that locus and then examining for allelic association at that site with the disease.

2. Complete genomic mapping: This is based on the production of complementary DNA (c.DNA) and genomic DNA (g.DNA) libraries, which allow the isolation of random unique DNA fragments with regular spacing along each and every chromosome and the subsequent use of these fragments as hybridisation probes to detect polymorphism within the genome. Pedigree studies may then be pursued using these alleles and should any disease association be found the gene fragment can be mapped to the genome and the underlying aetiological mutation tracked down by "walking" along the chromosomal segment with further probes.[7]

GENETICS OF ATHEROSCLEROSIS

The identification of candidate genes for atherosclerosis has thus been based on data regarding proteins thought to be implicated in atherogenesis. Examples of such include the apolipoproteins, the LDL receptor and many others. A list of such candidate genes is presented in table 1. Some of those showing protein polymorphism have already been studied with regard to associations with atherosclerosis.

Protein Polymorphism

Apolipoprotein E

This polymorphism has three common alleles known as E2, E3 and E4 (table 2) and a series of rarer alleles more often found in patients with type III hyperlipidaemia.[8]

The three common isoforms are known to differ by specific amino-acid replacements in two positions of the peptide chain and thus by their functional properties. For example apo E2 (arg-158 to cys) is defective in

Table 1. Candidate Genes for Atherosclerosis

Phenotype	Protein	Chromosomal Location	PIC
Lipoproteins	Apolipoproteins		
	AI-CIII-AIV	11q23-24	.73/.36/.55
	E-CI-CII	19q13	.36/.28/.79
	B	2p24-23	.66
Receptors	LDL receptor	19p13	.60
	Remnant receptor	-	-
	Insulin receptor	19p13	.80
Enzymes	LCAT	16q22	-
	Lipoprotein Lipase	8p22	.57
Vessel/wall proteins	Fibronectin	2q34-36	.36
	Collagen	17q21-22	.43
Growth Factors	PDGF B	22q12-13	.37
	PDGF A	7p21-p22 or 7q11-12	-
	Epidermal GF	-	-
	Insulin	11p15	.57
Coagulation Factors	Fibrinogen A	4q28	.34
	Fibrinogen B	4q28	.28
	Prothrombin	-	-
	Factor VII	13q34	-

Table 2. Apo E Polymorphisms

Apo E Alleles	Protein	Polymorphism
E2	Apo E2 (arg-158 to cys)	Receptor binding activity <2% of apo E3
E3	Apo E3	-
E4	Apo E4 (cys-112 to arg)	Enhanced in vivo catabolism

Frequencies of Alleles of Apo E

Apo E Alleles	Finns (n=408)	Germans (n=1031)	Japanese (n=319)
E2	0.029	0.077	0.081
E3	0.750	0.773	0.849
E4	0.221	0.150	0.067

binding to its lipoprotein receptors and thus, when homozygous predisposes to type III hyperlipidaemia. Most E2/E2 subjects, however, never develop hyperlipidaemia, but on the contrary have subnormal plasma cholesterol levels (mean effect of -0.367 mmol/l) due to reduced concentrations of LDL. Conversely subjects with the E4 allele have a raised plasma cholesterol (+0.181 mmol/l). The apo E gene locus is claimed to account for 4% and 20% of the phenotypic variance of plasma cholesterol and apo E concentrations respectively in German populations.[9] This is reflected in the differing levels of LDL cholesterol found amongst survivors of myocardial infarction with differing apo E phenotypes (apo E3/4 5.15 mmol/l $\pm$ 1.16 and apo E2/3 4.21 mmol/l $\pm$ 1.29).[10] Another study has examined polymorphism of apo E in relation to the risk of premature CAD.[11] The isotype frequencies were determined in a random sample of 400 people, aged 45 to 60 years, living in NE Scotland and compared to those found in a group of survivors of myocardial infarction, aged less than 56 years collected from diverse sources. The isotype mix E4/3 was seen more frequently in the CAD group than in the controls at the expense of E3/2 (0.32 vs 0.25 and 0.075 vs 0.127) and for survivors aged under 60 years this heterogeneity was even more marked.

Comparison of the average age of first myocardial infarction in male survivors also suggested that this may have occurred earlier in those of phenotype E4/3 (E3/3 53.95 $\pm$ 0.68, E4/3 51.20 $\pm$ 0.98, E3/2 53.21 $\pm$ 2.02). This does raise the possibility that the E4 allele may play a role in the aetiology of premature coronary heart disease but the frequency of alternative phenotypes amongst the survivors of myocardial infarction could also be explained as arising from a reduced survival of infarction in the subjects with alternative phenotypes.

Lp(a) lipoprotein

The Lp(a) antigen was originally described almost 30 years ago.[12] Essentially it appears to consist of an LDL particle in which the apo B100 moiety is linked by a disulphide bond to apo(a), a glycoprotein with some structural homology to plasminogen. Lp(a) is a heterogeneous molecule varying in size and density as the apo(a) moiety varies in size (from 280kD to 700kD). This heterogeneity is explicable in terms of varying numbers of the repeating "Kringle 4" domain comprising the bulk of the Lp(a) molecule. The structure of these domains can also vary to a slight degree. Such variation is encoded by the apo(a) gene, found on chromosome 6.[13] More than seven apo(a) isoforms have been described to date and their size appears to be inversely related to the Lp(a) concentration in the plasma.

The cysteine involved in the disulphide linkage is thought to be located in Kringle 36 of apo(a) but where this linkage occurs and the other steps in the assembly of apo(a) are not currently known. The liver, however, is known to represent the major source of Lp(a) in the plasma and Lp(a) phenotypes are known to change after liver transplantation. The mode of clearance of Lp(a) is also obscure, though it is thought to be at least in part dependent on the hepatic LDL receptor, as some workers have noted raised levels of Lp(a) in patients with familial hypercholesterolaemia (FH).

The molecular heterogeneity of Lp(a) and its similarity to plasminogen greatly complicates its effective assay, raising difficulties for clinical studies. Although the most reliable assay for Lp(a), the ELISA technique, appears to be free of interference by plasminogen, few clinical studies have yet made use of this. Studies using other less reliable methods of assay have shown that in case-control studies Lp(a) levels of more than approximately 20 mg/dl are associated with an increased risk of myocardial or cerebrovascular infarction.[14,15] A recent study has shown that individuals with a plasma concentration of greater than 25mg/dl exhibit a twofold higher risk of myocardial infarction than controls.[16]

As to the mechanism of this association, it has been suggested that Lp(a) may be affecting the fibrinolytic system and blocking the action of plasminogen thus predisposing to thrombogenesis. This possibility is suggested by the structural homologies between plasminogen and Lp(a). It should, however, be noted that apo (a) contains solely the Kringle 4 domain of plasminogen which is known to bind fibrin only weakly, unlike Kringle 1 domains. The issue is complicated by the existence of many variant forms of Kringle 4 within Lp(a) molecules and the varying numbers of these units. There is a strong possibility that the larger isoforms of Lp(a) may have differing effects from the smaller ones. All isoforms of Lp(a) need to be studied before the thrombogenic potential of Lp(a) can be assessed and this is a prerequisite to assessing the atherogenic role of the apolipoprotein.

Genetic Polymorphism

Complementing this data regarding protein polymorphism there is a parallel body of data regarding genetic polymorphism and susceptibility to atherosclerosis.

Apolipoprotein AI-CIII-AIV gene cluster

These three genes are congregated on the long arm of chromosome 11, covering a segment of DNA of approximately 15kb in length.[17,18] The organisation of the cluster shows the following features;

1. The apo CIII gene is transcribed in the opposite direction to the apo AI and AIV genes despite their proximity.

2. More than nine restriction enzyme dimorphisms occur within the cluster.[19,20] Many population studies have been performed examining the frequencies of alleles at these restriction sites to examine for association with premature CAD and its lipoprotein intermediates and the results are as follows;

United Kingdom: Two groups of patients have been studied , young survivors of myocardial infarction [21] and patients with coronary and extra-coronary atheroma demonstrated by angiography.[22] In the former group the frequency of an uncommon allele (the S2 allele) at the Sst I restriction site, within the fourth exon of the apo CIII gene, was approximately 4% in healthy controls (n=47) compared to 21% in young survivors of myocardial infarction (n=48). When other restriction sites polymorphisms were included in the analysis, thereby constructing DNA haplotypes,[23] it was found that one particular haplotype containing the uncommon allele at both the Msp I and Sst I site was increased from 2% in normolipaemic controls (n=48) to 21% in survivors of myocardial infarction (n=47) giving a relative incidence of 12.7 (p<0.01). It was not, however, possible to identify haplotypes with any greater association with premature CAD than the S2 allele alone. In a study from Edinburgh, [24] the Sst I site S2 allele, combined with rare alleles at the Xmn I, Pst I and Msp I sites showed an increased frequency in patients with coronary atherosclerosis who had a positive family history of the disease, as compared to those without a family history (relative incidence 3.34, p<0.0005).

Caution must, however, be exercised in attempting to interpret these results, as CAD is not a disease of homogeneous pathology, nor are such population groups free from ethnic heterogeneity. To achieve the best results the patient groups must be clearly defined and standardised as much as possible with regard to racial origin and clinical diagnostic features.

Since atherosclerosis has a variable age of onset, control groups may contain individuals who will later go on to develop the disease. For example the frequency of the S2 allele in the control group for the Scottish study was 18% (n=64) as compared to 4% for the London based studies (n=47). This may of course represent a real difference in allelic frequencies of the RFLP between ethnically differing populations but it may merely represent the differences in selection criteria applied to create the two groups (ie. presence or absence of hyperlipidaemia or other risk factors for coronary heart disease).

In the light of such considerations particular weight may be laid on studies providing for angiographic assessment in both patient and control groups. In one such study by Rees et al the frequency of the rarer S2 allele of the Sst I site was found to be 22% in patients with severe obstructive coronary atheroma (n=61) as compared to 6% in subjects with minimal disease (n=68,p<0.02).[22]

United States: Studies from Boston, Seattle and New York have been reported. In the first [25] caucasian patients (n=88) with severe CAD were compared to a Framingham control population (n=64) matched for ethnic origin and with other clinical criteria carefully standardised. The frequency of the uncommon allele revealed by the enzyme Pst I at a restriction site 34bp 3' to the apo AI gene was 32% in patients compared with 4% in the controls (p<0.01) and 3% in 30 subjects with no angiographic evidence of CAD, giving a relative risk of at least 10. The same rare allele was found at increased frequency in subjects with familial hypoalphalipoproteinaemia. Frequencies of alleles at other polymorphic sites at this locus were not reported. In Seattle, however, frequencies of alleles revealed by Pst I were found to be similar in random normal groups and patient groups with CAD defined by angiography (n=140).[26] The background frequency of the allele P2 was, however, markedly different from that reported in the Boston study. Clearly this kind of discrepancy would tend to minimise observable disease associations. Sst I alleles compared in the same Seattle study were found to differ significantly (p<0.05) between controls and patients (S2 allele;0.06 vs 0.12, n=101 and n=140 respectively).

West Germany: A study from Munster [27] examined eight polymorphic sites at the apo AI-CIII-AIV gene cluster. These included those of the restriction enzymes Apa I, Msp I, Pst I, Bam II and Pvu II. Pseudohaplotypes were constructed for 314 patients suffering from premature CAD (myocardial infarct before age 45) and compared with those of 267 student controls. Given considerations of the age of expression of the disease phenotype in atherosclerosis and the importance of group homogeneity the similar allelic frequencies reported in the respective study groups were neither unexpected nor informative.

Japan: A study from Northern Japan compared 69 subjects surviving myocardial infarction with 82 controls. The haplotype S1-M2 was significantly increased in the patient group (0.24 vs 0.11, p<0.05). As might be expected control frequencies differed markedly from caucasoid norms.

All the studies examined here apart from the Munster study, seem to support the hypothesis that within the apo AI-CIII-AIV gene cluster there exists an aetiological locus for atherosclerosis which accounts for several linkage disequilibrium phenomena observeable with different RFLP's in separate and unique populations. This linkage disequilibrium is illustrated for the most studied RFLP, the Sst I site, in table 3. The varying degree of association observed would be in keeping with a situation in which linkage with a separate aetiological mutation underlies the reported associations. The association of differing linkage markers for equivalent aetiological mutations occurring in differing populations has been well documented in other situations.[28]

Table 3. Sst I Allele Frequencies

Allelic Frequencies

Control Groups	n	S1	S2	References
Random medical outpatients	37	0.96	0.04	Rees et al
Health screen clinics				
Samples 1	42	1.0	0.00	Rees et al
2	74	0.98	0.02	Ferns et al
3	56	0.98	0.02	O'Connor et al
Normal coronary arteries	68	0.97	0.03	Rees et al
Random medical outpatients	35	0.99	0.01	Trembath et al
Normolipidaemic controls	71	0.94	0.06	Kessling et al
Random normals	101	0.94	0.06	Deeb et al
Controls	66	0.98	0.02	Hegele et al

Patient Groups	n	S1	S2	References
Hyperlipidaemic (iv/v)	28	0.80	0.20	Rees et al
Survivors of MI	48	0.88	0.12	Ferns et al
Coronary atheroma	61	0.89	0.11	Rees et al
Peripheral atheroma	49	0.88	0.12	O'Connor et al
Diabetic survivors of MI	47	0.86	0.14	Trembath et al
Hyperlipidaemia with gout	22	0.88	0.12	Ferns et al
Coronary heart disease	140	0.88	0.12	Deeb et al
Survivors of MI	66	0.96	0.04	Hegele et al

Low Density Lipoprotein (LDL) receptor gene

The elevation in LDL cholesterol resulting from familial hypercholesterolaemia (FH) is known to lead to premature CAD and this may account for up to 6% of myocardial infarction occurring before the age of 60 years.[29] The human LDL receptor gene has been cloned and localised to chromosome 19 and shown to consist of 18 exons, 13 of which have marked sequence homology to the genes for the C9 component of complement and for epidermal growth factor.[30,31]

RFLP studies at this locus have demonstrated segregation with FH in several isolated families [32,33] and within certain well characterised hypercholesterolaemic groups the defect underlying LDL receptor failure is known to be genetic and has been elucidated.[34,35,36] In certain instances this knowledge has led to homozygous FH patients as being identified as possessing either similar or dissimilar gene defects on each parentally derived chromosome. Single gene defects can be often shown to account for high levels of FH present in populations with marked founder effects, such as the Quebecois, the Lebanese and the Afrikaaners.

The apolipoprotein B gene

The gene for human apo B has been cloned and localised to chromosome 2 in the region p24.[37,38] It extends over 43kb containing 29 exons and 28 introns. The distribution of these introns is somewhat asymmetrical with most occurring in the 5'-terminal third of the gene. The sequence of the coding portion of the gene is known and the protein structure has been deduced from this.[39] A domain rich in basic amino-acids has been identified as important for the cellular uptake of cholesterol by the LDL receptor pathway. Many

RFLP's have been observed at the locus including those of the enzymes Xba I, Eco RI and Msp I and several studies have addressed the question of possible disease associations. In one such U.K. study the allelic frequencies of the Xba I polymorphism were not noted to be significantly different in 52 survivors of myocardial infarction and 33 healthy controls.[40] This was also noted in a US study where very similar allele frequencies were reported.[26] In this latter study there was, however, found to be a change in the frequency of an Msp I revealed insertional-deletional polymorphism, which increased in frequency in patient groups (from 0.06 to 0.15; n=62 and n=103). This finding was supported in another study where the frequency of the insertional polymorphism was increased from 0.142 in controls to 0.267 in patients (n=84 and n=84).[41] In addition the latter study reported an increased frequency of the rarer allele of the Xba I polymorphism in patients compared to controls. The authors concluded that such polymorphisms were acting as genetic markers in linkage disequilibrium with aetiological mutations nearby. None of the RFLP's were associated with any intermediate phenotypes of atherosclerosis such as lipoprotein or apolipoprotein levels. The small fluctuations of allelic frequencies seen across the board for such linkage markers would seem to suggest a role for the apo B gene as an intermediate or minor gene for the development of premature CAD.

Lipoprotein Lipase (LPL)

The gene for human lipoprotein lipase has been cloned and localised to chromosome 8 in the region p22 [42] and its gene product is known to play a central role in lipid catabolism, catalysing the rate limiting step in the removal of triglyceride rich particles from the plasma. Two RFLP studies at this locus [43,44] have demonstrated a promising association with CAD for a polymorphic Hind III site thought to lie between exons 8 and 9 at the 3' end of the gene. The H2 allele associating with premature CAD would also appear to associate with the intermediate phenotype of hypertriglyceridaemia in at least two separate populations (Caucasian and Japanese).[45] The Pvu II site of LPL has been similarly associated with the variation of fasting plasma triglyceride in a random U.K. Caucasian population.[45]

Although an excess of triglyceride rich lipoproteins may not be an independent risk factor from HDL for the development of coronary atherosclerosis, the close metabolic inter-relationship between HDL_2 and triglyceride makes it very difficult to evaluate their separate roles. Hypertriglyceridaemia with low HDL, would seem to constitute an important risk factor for the development of coronary atherosclerosis and this association of linkage markers at the LPL gene and both hypertriglyceridaemia and CAD would seem to suggest that this gene might be a causal determinant of atherosclerosis, though this has yet to be confirmed and its relative importance assessed.

CONCLUSION

It would thus seem that atherosclerosis is an archetypal multifactorial disease with a strong genetic component. As the techniques of recombinant DNA technology are increasingly applied to its analysis the roles of linkage markers for the disease and intermediate lipoprotein phenotypes will hopefully become correspondingly well understood. A handful of genes, mostly those coding for the apolipoproteins and proteins associated with the metabolism of such apolipoproteins, have already been identified as minor or intermediate genes in the multigenic background of atherosclerosis, these include the apo E gene, the apo AI-CIII-AIV gene cluster, the LDL-receptor gene and the apo B gene. There remains, however, a great deal more of this genetic background to be discovered in the search for possible major aetiological genes for premature atherosclerosis and CAD.

REFERENCES

1. W. M. Yater, A. H. Traum, W. G. Brown, R. P. Fitzgerald, M. A. Geisler and B. B. Wilcox, Coronary artery disease in men eighteen to thirty-nine years of age, Am Heart J, 36:334 (1948).
2. J. Slack and K. A. Evans, The increased risk of death from ischaemic heart disease in first-degree relatives of 121 men and 96 women with ischaemic heart disease, J Med Genet, 3:239 (1966).
3. A. M. Rissanen, Familial aggregation of coronary heart disease in a high incidence area (North Karelia, Finland), Br Heart J, 42:294 (1979).
4. J. J. Nora, R. M. Lortscher, R. D. Spangler and W. J. Kimberling, Genetic epidemiology study of early onset ischaemic heart disease, Circulation, 61:503 (1980).
5. K. Berg, Twin studies of coronary heart disease and its risk factors, Acta Genet Med Gemell, 33:349 (1984).
6. K. Berg, Genetics of coronary heart disease, Prog Med Genet, 5:36 (1983).
7. S. M. Wiessman, Molecular genetic techniques for mapping the human genome, Mol Biol Med, 4:133 (1987).
8. G. Utermann, M. Hees and A. Steinmetz, Polymorphism of apoliporotein E and occurrence of dysbetalipoproteinaemia in man, Nature, 269:604 (1977).
9. G. Utermann, Apolipoproteins, quantitative liporotein traits and multifactorial hyperlipidaemia, CIBA Foundation Symposium, 130:52 (1987).
10. G. Assmann, H. Schulte, H. Funke, G. Schmitz and H. Robenck, High density lipoproteins and atherosclerosis. Atherosclerosis, 75:341 (1989).
11. A. N. Cumming and F. W. Robertson, Polymorphism at the apolipoprotein E locus in relation to risk of coronary disease, Clin Genet, 25:310 (1984).
12. K. Berg, A new serum type system in man: The Lp- system, Acta Pathol Microbiol Scand, 59:369 (1963).
13. J. W. McClean, J. E. Tomlinson, W. J. Kuang et al, cDNA sequence of human apolipoprotein (a) is homologous to plasminogen, Nature, 330:132 (1987).
14. G. M. Kostner, P. Avogaro, G. Gazzolato, E. Marth and G. Bittolobon, Lipoprotein Lp(a) and the risk of myocardial infarction, Atherosclerosis, 38:51 (1981).
15. V. W. Armstrong, A. K. Walli and D. Sendel, Isolation, characterization and uptake in human fibroblasts of an apo(a) free lipoprotein obtained on reduction of lipoprotein (a), J Lipid Res, 26:1314 (1985).
16. G. M. Kostner, Lipoprotein Lp(a) and HMG-CoA reductase inhibitors, Atherosclerosis, 75:405 (1989).
17. S. K. Karathanasis, V. I. Zanthis and J. L. Breslow, Linkage of human apolipoprotein AI and CIII genes, Nature, 304:371 (1983).
18. S. K. Karathanasis, Apolipoprotein multigene family: tandem organisation of human apoliporotein AI, CIII and AIV genes, Proc Natl Acad Sci USA, 82:6374 (1985).
19. A. Rees, J. Stocks, C. C. Shoulders, D. J. Galton and F. E. Baralle, DNA polymorphism adjacent to the human apoprotein AI gene in relation to hypertriglyceridaemia, Lancet, i:444 (1983).
20. J. J. Seilhammer, A. A. Protter, P. Frossard and B. Levy-Wilson, Isolation and DNA sequence of full length cDNA of the entire gene for human apolipoprotein AI. Discovery of a new polymorphism, DNA, 3:309 (1984).
21. G. A. A. Ferns, J. Stocks, C. Ritchie and D. J. Galton, Genetic polymorphisms of apolipoprotein C-II and insulin in survivors of myocardial infarction, Lancet, i:300 (1985).

22. A. Rees, N. I. Jowett, L. G. Williams et al, DNA polymorphisms flanking the insulin and apolipoprotein CIII genes and atherosclerosis, Atherosclerosis, 58:269 (1985).
23. G. A. A. Ferns and D. J. Galton, Haplotypes of the human apoprotein AI-CIII-AIV gene cluster in coronary atherosclerosis, Hum Genet, 73:245 (1986).
24. W. H. Price, S. W. Morris, A. H. Kitchin, P. R. Wenham, P. R. S. Burgon and P. M. Donald, DNA restriction fragment length polymorphisms as markers of familial coronary artery disease, Lancet, i:407 (1989).
25. J. M. Ordovas, E. J. Schaeffer, D. Salem et al, Apolipoprotein A-I gene polymorphism associated with premature coronary artery disease and familial hypoalphalipoproteinaemia, New Engl J Med, 314:671 (1986).
26. S. Deeb, A. Failor, B. G. Brown, J. D. Brunzell, J. J. Albers and A. G. Motulsky, Molecular genetics of apolipoproteins and coronary heart disease, Cold Spring Harb Symp Quant Biol, 51:403 (1987).
27. G. Assman, H. Schulte, H. Funke, G. Schmitz and H. Robenck, High Density Lipoproteins and atherosclerosis, Atherosclerosis, 75:34 (1989).
28. S. E. Humphries, L. G. Williams, A. F. Stahenhoef et al, Familial apolipoprotein CII deficiency: a preliminary analysis of the gene defect in two pedigrees, Hum Genet, 65:151 (1984).
29. J. L. Goldstein, W. R. Hazzard and H. G. Schrott, Hyperlipidaemia in coronary heart disease. Lipid levels in 500 survivors of myocardial infarction, J Clin Invest, 52:1533 (1973).
30. T. C. Sudhof, J. L. Goldstein, M. S. Brown and D. W. Russell, The LDL receptor gene : a mosaic of exons shared with different proteins, Science, 228:815 (1985).
31. U. Francke, M. S. Brown and J. L. Goldstein, Assignment of the human gene for the low density lipoprotein receptor to chromosome 19: Synteny of a receptor, a ligand and a genetic disease, Proc Natl Acad Sci USA, 81:2826 (1986).
32. S. E. Humphries, B. Hortshemke, M. Seed M et al, A common DNA polymorphism of the LDL receptor gene and its use in diagnosis, Lancet, i:1003 (1985).
33. M. F. Leppert, S. J. Hasstedt, T. Holm et al, A DNA probe for the LDL receptor gene is tightly linked to hypercholesterolaemia in a pedigree with early coronary disease, Am J Hum Genet, 39:300 (1986).
34. H. Tolleshaug, K. K. Hobgood, M. S. Brown and J. L. Goldstein, The LDL receptor locus in familial hypercholesterolaemia. Multiple mutations disrupt the transport and processing of a membrane receptor, Cell, 32:941 (1983).
35. M. A. Lehrmann, W. J. Schneider, T. C. Sudhof, M. S. Brown, J. L. Goldstein and D. W. Russell, Mutation in LDL receptor: Alu-Alu recombination deletes exons encoding transmembrane and cytoplasmic domains, Science, 227:140 (1985).
36. B. Horsthemke, A. M. Kessling, M. Seed, V. Wynn, R. Williamson and S. E. Humphries, Identification of a deletion in the low density lipoprotein (LDL) receptor gene in a patient with familial hypercholesterolaemia, Hum Genet, 71:75 (1985).
37. J. J. Knott, S. C. Rall Jnr, T. L. Innerarity et al, Human apolipoprotein B: structure of carboxy-terminal domain sites of gene expression and chromosomal localisation, Science, 230:37 (1985).
38. C. C. Shoulders, N. Myant, A. Sidoli, J. C. Rodriguez, C. Cortese and F. E. Baralle, Molecular cloning of human LDL apolipoprotein B cDNA, Atherosclerosis, 58:277 (1985).
39. T. J. Knott, R. J. Pease, L. M. Powell et al, Human apolipoprotein B: complete cDNA sequence and identification of domains of the protein, Nature, 323:734 (1986).
40. G. A. A. Ferns and D. J. Galton, Frequency of the Xba I polymorphisms of the apolipoprotein B gene in myocardial infarct survivors, Lancet, ii:572 (1986).
41. R. A. Hegele, L. S. Huang, P. N. Herbert et al, Apolipoprotein B-gene DNA

polymorphisms associated with myocardial infarction, <u>New Engl J Med</u>, 315:1509 (1986).

42. R. S. Sparkes, S. Zollman, I. Klisak et al, Human genes involved in lipolysis of plasma lipoproteins : mapping of loci for lipoprotein lipase to 8p22 and hepatic lipase to 15q21, <u>Genomics</u>, 1(ii):6763 (1987).

43. J. Thorn, J. C. Chamberlain, J. Stocks and D. J. Galton, RFLP's at the lipoprotein lipase and hepatic lipase gene loci in coronary atherosclerosis, <u>Atherosclerosis</u>, 79:94 (1989).

44. J. C. Chamberlain, J. A. Thorn, R. Morgan et al, A linkage marker at the human lipoprotein lipase gene locus associates with coronary atherosclerosis in a welsh population. Submitted for publication.

45. J. C. Chamberlain, J. A. Thorn, K. Oka, D. J. Galton and J. Stocks, DNA polymorphisms at the lipoprotein lipase gene: associations in normal and hypertriglyceridaemic subjects, <u>Atherosclerosis</u>, 79;85 (1989).

RFLP MARKERS OF FAMILIAL CORONARY HEART DISEASE

William H. Price and Arthur H. Kitchin

University Department of Medicine and M.R.C. Human Genetics Unit,
Western General Hospital, Edinburgh EH4 2XU

Abstract

In two population samples of men aged 30-59 years from the North of Edinburgh, a higher frequency of CHD in those with a family history of the disease in close relatives before age 60 was strongly associated with the allele status of four RFLP in the apo AI/CIII/AIV gene region. In a simultaneous multiple logistic regression analysis of CHD on a number of covariates, a family history emerged with significant odds ratio of 2.553 (p = 0.034) in subjects with minor RFLP alleles in one of the samples and with odds ratio of 3.611 (p = 0.016) in those homozygous for major alleles in the other. The apo AI/CIII/AIV related effect of family history on evidence of CHD was independent of all other variables tested, which included age, cigarette smoking, hypertension, hyperlipidaemia, socio-economic group and obesity. There was no evidence of a familial environment effect or of genotype environment interaction. The results of these studies suggest that genetic liability to CHD is influenced significantly by a locus or loci in the apo AI/CIII/AIV gene region. Since the association of this region with familial CHD involves different alleles or haplotypes in different sections of the population (different linkage phase) no single allele or haplotype identifies an individual with increased liability to CHD on account of this association. This may explain the apparently conflicting results obtained in case control studies.

Introduction

A number of factors are known to correlate with an increased risk of developing coronary heart disease (CHD). They include cigarette smoking, an elevated blood pressure, raised levels of total and LDL cholesterol and of apo-B, reduced levels of HDL cholesterol and apo-AI. There may be an association with raised plasma triglyceride levels as well but it is not yet clear if this is independent of cholesterol levels. There are CHD associations with obesity and low physical activity but they are relatively slight and likely to be indirect. In some communities, the incidence of the disease appears to correlate with indices of poverty and unemployment, both of which have a strong influence on mortality from all causes. The composition of the drinking water supply in some areas may also contribute. Some of the major risk factors have a familial component. This may be genetically determined as in some cases of familial hyperlipidaemia or it may be due to habits learned in a shared home environment during childhood and adolescence as in the case of cigarette smoking and diet dependent hyperlipidaemia. But there is also very strong evidence of a familial component that is independent of other risk factors. Thus it has been shown in many studies that the first degree relatives of young CHD patients are at increased risk of developing the disease prematurely even after other factors have been taken into account (1, 2, 3). The effect is age and sex related since in most studies it has been observed only in men and generally before the age of 60. The relative contributions of genetic and environmental factors have been extensively investigated and there is considerable evidence of an important genetic contribution. In recent years, efforts have been directed towards identifying the individual genes that might be implicated. They make use of new developments in molecular genetics and the discovery of restriction fragment length polymorphisms (RFLP) as genetic markers. Because of the early development of DNA

DNA Polymorphisms as Disease Markers, Edited by D.J. Galton and
G. Asmann, Plenum Press, New York, 1991

probes for the apo AI/CIII/AIV gene cluster and the involvement of these genes in the control of lipid metabolism, many of the earlier studies concentrated on this region. Added impetus was provided by the report of a family with a deletion/insertion rearrangement involving the apo AI/CIII genes. It resulted in very low or absent plasma HDL-cholesterol and apo-AI and very severe cardiovascular disease (4). Subsequently, an association was described between the minor allele at a SacI (SstI) RFLP at the 3' end of the CIII gene and hypertriglyceridaemia (5) and shortly afterwards between the same minor allele and premature CHD (6). There followed a spate of reports confirming or refuting these observations and some describing CHD association with other RFLP in the same region (7, 8). The inconsistency that characterised these studies has since been observed with other ''candidate'' genes such as the apo-B gene. When dealing with a disease such as CHD where the cause is multifactorial and the genetic component probably polygenic there are a number of reasons why association with polymorphic markers such as RFLP should vary between populations (9). Firstly, the presence of a predisposing gene at one locus may have no phenotypic effect if genes at other loci are not also of the predisposing type. Secondly, the effect of a gene may be dependent on an appropriate environment. Thirdly, linkage disequilibrium with an RFLP may not involve the same RFLP allele in different populations. Fourthly, the strength of the linkage disequilibrium may vary between populations. Another source of apparent discrepancy is the use of different criteria for diagnosis of coronary heart disease in different studies.

In the North of Edinburgh we have investigated the possible association of CHD and RFLP in the region of the apo AI/CIII/AIV gene in two samples of the population. One of these was from a relatively stable and well established area which was formerly the ancient burgh and port of Leith and the other from a socially and geographically distinct area in a more recently developed suburban residential district of Cramond. In both we found significant associations with familial CHD (10, 11) i.e. CHD attributable to a history of premature disease in at least one first degree relative, when apo AI/CIII/AIV RFLP status was taken into account. In this paper we describe the independence of this association from other CHD risk factors most of which occur with very different frequencies in the two samples.

In order to estimate the likely frequency of CHD in the two population samples, we examined mortality from this cause in the two areas over a 14 year period. From the 1981 population census, the numbers of males and their distribution by age groups was obtained for the postcode sectors of North Edinburgh. The male death rate from ischaemic heart disease in Scotland (Registrar General for Scotland) was averaged for each age group over the 14 year period and applied to the census data to give expected annual death rates for men in all the postcode sectors represented in the population samples. Actual male deaths from IHD in each postcode sector were available for 1974-87. These were averaged over the period and the resulting Standardised Mortality Ratio (actual/expected deaths) from IHD was calculated. This varied from 0.53 to 1.27 in different postcode sectors. An SMR weighted according to the postcodes of participants in the study was calculated for each of the two samples. This figure was 0.98 in Leith and 0.77 in Cramond.

Methods

In both samples, the subjects studied were men aged 30-59 years. They had all been invited to participate in the study by letter from their general medical practitioner. Respondents to the GP's letter were asked to attend a clinic where a WHO cardiovascular questionnaire and a Scottish Heart Health Study Family History questionnaire were completed, height and weight recorded and blood pressure measured using the Random Zero London School of Hygiene instrument. A 12 lead electrocardiogram was recorded. Evidence of the diagnosis of CHD was based on the questionnaire responses and the ECG findings classified by the Minnesota Code. Blood was collected for haemoglobin and measurement of carbon monoxide by cooximetry, and the following lipid analyses were carried out: plasma total cholesterol by an enzymatic (cholesterol oxidase) method, triglyceride by an enzymatic (glycerol kinase) method, HDL cholesterol by phosphotungstate precipitation, apo AI, AII and B by immunoturbidimetry. LDL cholesterol was calculated from the formula LDL = TC − (HDL + TG / 2.19). Subjects whose non-fasting total cholesterol (TC) exceeded 6.4 mmol/l or triglyceride (TG) 2.2 mmol/l were recalled for a fasting blood sample and measurements were repeated.

For RFLP analysis, 15mls of blood were collected into mucus heparin and DNA was extracted by the method of Kunkel et al (12). The allele status at four RFLP identified with the enzymes XmnI, MspI, PstI and SacI in the region of the apo AI/CIII genes was determined by Southern analysis as previously described (11).

Table 1. Age distribution

Age (years)	Leith n (%)	Cramond n (%)
30-34	62 (8.7)	54 (6.2)
35-39	130 (18.2)	121 (14.0)
40-44	107 (15.0)	252 (23.3)
45-49	147 (20.6)	141 (16.3)
50-54	127 (17.8)	174 (20.1)
55-59	140 (19.6)	174 (20.1)
Total	713 (100)	866 (100)

Statistical Analysis

Frequencies were compared by χ^2 analysis or the exact probability test when numbers were small. Quantitative variables were compared by the student t test or the non-parametric Mann-Whitney U test or Wilcoxson's rank sum test. Correlation of lipid values by age were assessed by analysis of variance. The independent contributions of multiple variables were assessed simultaneously by stepwise logistic regression analyses.

Results

The response rate of 1059 subjects in Leith was 67% (713) and of 1239 subjects in Cramond was 71% (879). In Leith 84% of respondents and 80% in Cramond had Scottish surnames. Of the Leith men 72% had been born in Edinburgh compared with 33% in Cramond.

The age distribution and socio-economic status of respondents are shown in tables 1 and 2. The Cramond subjects are slightly older ($\chi^2 = 26.65$ p < 0.001). The distribution by social group was markedly different ($\chi^2 = 418.82$ p < 0.001) in the populations. In Cramond 70% were professional workers or generally well to do (groups I and II) and less than 3% were unskilled workers (groups IV and V). In Leith less than 25% belonged to groups I and II and over 75% were skilled artisans (group III) or unskilled labourers. Hypertension, obesity, smoking and alcohol intake are compared in the two populations in table 3. Hypertension defined as a systolic pressure > 160mm Hg and/or a diastolic pressure > 95mm Hg was present in 8.9% of all men studied with no significant difference between the two populations. Obesity, measured as weight / height^2 > 24.5 kg/m^2 was more common in Leith (57.3%) than in Cramond (42.7%) ($\chi^2 = 43.47$; p < 0.0001). There were more cigarette smokers ($\chi^2 = 228.87$; p < 0.0001) and alcohol consumption was markedly greater in Leith (p < 0.001).

Table 2. Social group distribution (Registrar General's classification)

Social group	Leith n (%)	Cramond n (%)
I	61 (8.9)	259 (29.9)
II	105 (15.3)	355 (41.0)
III	292 (42.5)	227 (26.2)
IV	162 (23.6)	19 (2.2)
V	67 (9.8)	5 (0.6)
Total	687 (100)	865 (100)

Table 3. Hypertension, obesity, smoking and alcohol intake

		Leith n (%)	Cramond n (%)	p
Blood pressure	Normal	643 (90.2)	796 (91.9)	
	Hypertensive	70 (9.8)	70 (8.1)	NS
	Total	713 (100)	866 (100)	
Weight / Height^2	Normal	302 (42.6)	495 (57.2)	
	Overweight*	325 (45.8)	325 (37.5)	
	Obese*	82 (11.5)	45 (5.2)	< 0.001
	Total	709 (100)	865 (100)	
Smoking	Non–smoker	182 (25.5)	447 (51.6)	
	Cigar / pipe	58 (8.1)	10 (1.2)	
	ex–/10	217 (30.4)	317 (36.6)	
	11–20	148 (20.8)	53 (6.1)	< 0.001
	21–40	100 (14.0)	36 (4.2)	
	40	8 (1.1)	3 (0.3)	
	Total	713 (100)	866 (100)	
Alcohol	0–20	462 (67.4)	730 (84.3)	
	21–36	122 (17.8)	94 (10.9)	
	37–50	61 (8.9)	36 (4.2)	
	51–95	34 (5.0)	5 (0.6)	< 0.001
	96	6 (0.9)	0 (0)	
	Total	685 (100)	865 (100)	

* Overweight = 24.5 kg/m^2 to 26.4 kg/m^2; Obese > 26.4 kg/m^2

Lipid data collection (table 4) was virtually complete except for some early subjects in the Leith sample in whom estimations of HDL (187) apo-AII (106) apo-B (69) and apo-AI (48) were omitted. Cholesterol, triglyceride and apo-AII were significantly higher in the Leith group while HDL and apo-AI were markedly lower. There was no significant difference in LDL cholesterol or apo-B between the two populations. The numbers recalled for fasting samples because of high cholesterol or triglyceride values were 329 / 713 in Leith (46.1%) and 272 / 830 (32.8%) in Cramond. Hyperlipidaemia defined as fasting total cholesterol exceeding 6.4 mmol/l or triglyceride exceeding 2.2 mmol/l was present in 271 (37.6%) of the Leith subjects and 234 (27.0%) of the Cramond subjects, a significant difference (p < 0.001). Thus Leith had both an excess of high fasting triglycerides and also an excess of temporarily elevated triglycerides which reverted to normal on fasting.

Table 4. Lipid measurements

	n	Leith Mean (SD)	n	Cramond Mean (SD)	p
Total cholesterol	713	6.02 (1.20)	780	5.88 (1.12)	= 0.018
Triglyceride	712	2.44 (1.57)	786	1.80 (1.01)	< 0.001
" fasting	329	2.12 (1.22)	272	1.80 (1.09)	< 0.001
HDL cholesterol	526	1.20 (0.44)	760	1.31 (0.30)	< 0.05
LDL cholesterol	525	3.70 (1.10)	631	3.77 (1.03)	= 0.237
Total / HDL chol	526	5.84 (3.73)	760	4.70 (1.49)	< 0.001
Apolipoprotein AI	665	1.04 (0.19)	824	1.20 (0.18)	< 0.001
AII	607	0.39 (0.13)	832	0.38 (0.09)	= 0.007
B	644	1.37 (0.37)	823	1.35 (0.36)	= 0.214
Apo B / AI	644	1.37 (0.46)	823	1.16 (0.37)	< 0.001

Table 5. Evidence of coronary heart disease

		Leith n (%)	Cramond n (%)
Questionnaire	Negative	653 (92.2)	801 (92.6)
	Mild angina	15 (2.1)	17 (2.0)
	Severe angina	23 (3.2)	38 (4.4)
	Possible infarct	17 (2.4)	9 (1.0)
	Total	708 (100)	865 (100)
Electrocardiogram	Normal	610 (85.6)	721 (84.6)
	Possible IHD	25 (3.5)	19 (2.2)
	Probable IHD	14 (2.0)	6 (0.7)
	Other abnormality	64 (9.0)	105 (12.3)
	Total	713 (100)	851 (100)
Diagnosis of CHD	Questionnaire only	43 (6.0)	59 (6.9)
	Cardiogram only	27 (3.8)	20 (2.3)
	Both	12 (1.7)	5 (0.6)
	Total	82 (11.6)	84 (9.9)

Evidence of coronary heart disease (table 5) was present in 10.6% of subjects overall. This was based on the questionnaire only in 6.5%, on the cardiograph only in 3.0% and on both in 1.1%. Differences between the two populations were not significant as regards the questionnaire, but cardiographic evidence of CHD was significantly more common in Leith than in Cramond ($p = 0.017$). Other cardiographic abnormalities, not related to CHD, were present in 10.8% of subjects.

The frequency of CHD in first degree relatives is compared in the two populations in table 6. A history of CHD in one or more first degree relatives before the age of 60 was provided by 34.5% of subjects in the two samples. The male parent was more commonly affected in both communities. Involvement of the mother was much more common in Leith than in Cramond (15% and 8%) and both parents were also more commonly involved in Leith. A history of CHD in sibs was obtained in 14.6% of subjects in Leith and 4.4% in Cramond. Sibs alone were involved in 7.6% and 2.3% respectively; male sibs were involved more frequently. There was a striking difference between the two communities in sibship composition, the average number of sibs in Leith being 3.12 and in Cramond 1.63 (table 7) and partly for this reason there was higher reporting of CHD in sibs in Leith than in Cramond.

Table 6. CHD in first degree relatives

	Leith		Cramond	
	n	(%)	n	(%)
No family history	427	(59.9)	604	(70.0)
Parental	232	(32.5)	239	(27.7)
Both parents	26	(3.7)	15	(1.7)
Father	151	(21.2)	187	(21.7)
Mother	107	(15.0)	67	(7.8)
Sibs:				
male	66 104	(14.6)	29 38	(4.4)
female	38		9	
Sibs only	54	(7.6)	20	(2.3)
corrected*	17.3	(2.4)	12.3	(1.4)
Positive family history	286	(40.1)	259	(30.0)
No of subjects	713	(100)	863	(100)

Table 7. Number of sibs

	Leith n (%)	Cramond n (%)	Total n (%)
0	62 (8.7)	153 (17.7)	215 (13.6)
1	161 (22.6)	344 (39.7)	505 (32.0)
2	141 (19.8)	198 (22.9)	339 (21.5)
3	100 (14.0)	96 (11.1)	196 (12.4)
4-6	168 (23.3)	64 (7.4)	232 (14.7)
7+	80 (11.0)	11 (1.2)	91 (5.8)
total	712 (100)	866 (100)	1578 (100)

Correcting for number of sibs gave a ratio of 1.5 for sib CHD in Leith compared with Cramond. The larger family size in Leith and more frequent maternal involvement account for the fact that a positive family history in first degree relatives was significantly more common in Leith (40.1%) than in Cramond (30.0%). A history of CHD in two or more first degree relatives was provided by 6% of subjects.

A parental history of CHD was unrelated to the number of sibs, while a history of CHD in sibs increased with the number of sibs (table 8).

The frequency of the minor alleles X_2, M_2, P_2 and S_2 identified with the enzyme XmnI, MspI, PstI and SacI did not differ significantly in the two populations. The distribution of the genotypes at each locus conformed with Hardy-Weinberg expectations both in Leith and Cramond (table 9). The RFLP alleles were distributed in four principal haplotypes (a) $X_1M_1P_1S_1$, (b) $X_2M_1P_1S_1$, (c) $X_1M_2P_1S_2$ and (d) $X_1M_1P_2S_1$. Other haplotypes were so few that they could not be analysed separately. In table 10 and elsewhere in the text the populations have been divided into four genotypes *(i)* major alleles only i.e. homozygous for haplotype a *(ii)* X_2, homozygous for haplotype b or heterozygous ab, *(iii)* P_2, homozygous for d or heterozygous ad and *(iv)* S_2M_2 homozygous for c or heterozygous ac (this category includes subjects who were homozygous or heterozygous for S_2 or M_2 alone). Subjects who were doubly heterozygous for haplotypes b, c and d were too few to categorise separately and data for these subjects are included with both genotypes when examing the effect of each minor haplotype separately, but in the category of "any one or more minor allele" these subjects were included once only.

Relationships Among Variables

With a few exceptions the same relationships were observed among variables in Leith and Cramond.

Age. Within the selected age range of 30-59 there were significant increases in the frequency of hypertension and obesity, and in the levels of cholesterol, LDL, apo-B and the ratio of apo B to AI. There was no significant increase in hyperlipidaemia, however, or in history of CHD, either personal or in a first degree relative.

Hypertension. This was significantly more common in the lower socio-economic groups and was also related to higher levels of triglyceride and of total and LDL cholesterol, to lower levels of HDL cholesterol and apo-AI, and raised apo-B / AI ratio.

Table 8. Family history of CHD and number of sibs

No of Sibs	0	1	2	3	4	5	6+
Subjects	213	504	339	196	112	72	139
Parental CHD%	32.9	28.8	26.3	32.7	33.0	31.9	30.2
Sib CHD%	-	3.2	5.9	14.3	10.7	13.9	30.9

Table 9. Restriction fragment length polymorphism - allele frequencies

	XmnI		MspI		PstI		SacI	
	Leith	Cramond	Leith	Cramond	Leith	Cramond	Leith	Cramond
Number of subjects	695	795	693	600	679	733	698	805
Minor allele frequencies	0.130	0.116	0.070	0.064	0.061	0.067	0.073	0.080
Genotype frequencies*								
pp	523 (525.8)	621 (621.3)	599 (598.5)	526 (525.7)	601 (598.5)	636 (638.1)	601 (599.7)	681 (681.8)
pq	163 (157.4)	163 (163.0)	90 (91.1)	71 (71.9)	73 (77.9)	96 (91.6)	92 (94.5)	120 (118.3)
qq	9 (11.8)	11 (10.6)	4 (3.5)	3 (2.5)	5 (2.5)	1 (3.3)	5 (3.7)	4 (4.8)

*Results shown as observed frequency (expected frequency from Hardy Weinberg equation).

Socio-economic group. In addition to a relation to hypertension, cigarette smokers, heavy alcohol drinkers and the obese were more frequent among the lower socio-economic groups.

Lipids. In both populations lipid interrelationships showed, as expected, strong correlations between total cholesterol levels and LDL cholesterol and apo-B (table 10). HDL cholesterol and apo-AI were also closely correlated and inversely related to total and LDL cholesterol, LDL and apo-B. The biggest difference in lipid levels between subjects with and without evidence of CHD was in apo-B and the ratio of B to AI but these differences did not achieve statistical significance. Non-fasting levels of triglyceride were related to obesity and alcohol intake. Hypertriglyceridaemia in fasting specimens of blood was related to hypertension, obesity, smoking, a family history of CHD, and in Leith to lower socio-economic status.

Coronary Heart Disease. Evidence of CHD revealed by questionnaire or cardiograph was present in 10.6% of subjects. In Leith, the frequency in those with a positive family history of CHD was 15.6% compared with 7.9% in those with no family history (p < 0.005). This increased frequency (table 11a) was confined to those with one of the minor RFLP alleles, X_2, M_2, P_2, S_2 (haplotypes b, c and d). In this group of subjects the frequency was 21.4% compared with 11.2% in those with no minor alleles (homozygous for haplotype a). In Cramond (table 11b), family history of CHD had very little effect on frequency of

Table 10. Lipid and apolipoprotein correlations (values of r)
Leith and Cramond populations combined

	Total chol	Trigly	HDL chol	LDL chol	Apo-AI	Apo-B	Apo-B / AI
Triglyceride	0.40						
HDL chol	NS	-0.42					
LDL chol	0.88	NS	-0.14				
Apo AI	0.12	-0.16	0.60	NS			
Apo B	0.77	0.48	-0.26	0.68	-0.09		
Apo B / AI	0.55	0.47	-0.50	0.53	-0.59	0.83	
LDL chol / HDL chol	0.51	0.21	-0.71	0.69	-0.33	0.55	0.59

NS = not significant p < 0.001 for all other figures

Table 11. Distribution of CHD morbidity by presence or absence of family history (FH+; FH−) and apolipoprotein AI/CIII RFLP genotype in (a) Leith (b) Cramond

(a)

AI/CIII/AIV RFLP	n	FH+ CHD (%)	n	FH− CHD (%)	p
HMZ for all four major alleles X_1, P_1, M_1, S_1	134	15 (11.2)	211	20 (9.5)	NS
HMZ or HTZ for minor alleles:					
X_2	54	11 (20.4)	113	6 (5.3)	< 0.05
P_2	27	6 (22.2)	49	2 (4.1)	< 0.05
M_2 and/or S_2*	36	8 (22.2)	76	6 (7.9)	< 0.05
Any one or more	103	22 (21.4)	219	14 (6.4	< 0.0005
Total	237	37 (15.6)	430	34 (7.9)	< 0.005

(b)

	n	CHD (%)	n	CHD (%)	p
HMZ for all four major alleles: X_1, P_1, M_1, S_1	92	16 (17.4)	269	24 (8.9)	< 0.05
HMZ or HTZ for minor alleles:					
X_2	51	2 (3.9)	118	8 (6.8)	NS
P_2	34	2 (5.9)	63	4 (6.3)	NS
M_2 and/or S_2*	41	5 (12.2)	89	11 (12.4)	NS
Total	212	24 (11.3)	521	47 (9.0)	NS

Values shown as number (%). HMZ = homozygous; HTZ = heterozygous; NS = not significant.

* Numbers of subjects with M_2 and S_2 are combined because these two alleles are in tight linkage disequilibrium and constitute a separate haplotype with P_1 and X_1.

personal CHD (11.3% compared with 9%) and was significant only in those who were homozygous for major RFLP alleles (i.e. haplotype a) at all 4 sites (17.4% vs 8.9%; p < 0.05). Subjects with evidence of CHD more frequently belonged to a lower socio-economic group. The frequencies of other factors in those with and without CHD are compared in table 12. Age and a family history are the only two which emerged as significant in either population.

Table 12. Comparison of those with (166) and without (1391) evidence of CHD in the combined population

	CHD	No CHD	p
Mean age (SD)	48.64 (8.13)	47.31 (7.98)	< 0.05
Mean weight / height^2 (SD)	25.40 (3.38)	25.37 (3.34)	NS
Family history CHD (%)	78 (47.0)	459 (33.0)	< 0.001
Smoking (%)	103 (62.0)	838 (60.1)	NS
Hypertensive (%)	19 (11.4)	116 (8.3)	NS
Lipids:			
Total cholesterol (SD)	6.02 (1.16)	5.94 (1.16)	NS
Total chol / HDL chol (SD)	5.63 (5.27)	5.13 (2.31)	NS
Triglyceride (SD)	2.18 (1.55)	2.08 (1.31)	NS
Apolipoprotein:			
AI (SD)	1.12 (0.24)	1.12 (0.19)	NS
B (SD)	1.41 (0.38)	1.35 (0.36)	NS
B / AI (SD)	1.31 (0.50)	1.24 (0.41)	NS

Table 13. Minor allele frequencies

Subjects	X_2 Leith	X_2 Cramond	M_2 Leith	M_2 Cramond	P_2 Leith	P_2 Cramond	S_2 Leith	S_2 Cramond
n	695	795	693	600	679	733	698	805
All men	0.130	0.116	0.070	0.064	0.061	0.067	0.073	0.080
CHD	0.115	0.075	0.078	0.086	0.059	0.055	0.069	0.094
TC above 6.4mmol/l	0.144	0.103	0.058	0.088	0.046	0.073	0.060	0.096
TG above 2.2mmol/l	0.123	0.090	0.062	0.115	0.062	0.033	0.080	0.118
Family history CHD	0.112	0.129	0.061	0.083	0.060	0.083	0.060	0.090

Family history of CHD. FH of CHD was related to social class, significantly lower levels of HDL, increased levels of total and LDL cholesterol, triglyceride, apo-B and apo-B / AI ratio as well as to CHD.

RFLP allele frequencies (table 13). No significant differences in minor allele frequencies were observed in subjects with CHD, total cholesterol above 6.4mmol/l, triglycerides above 2.2mmol/l or a family history of CHD. The frequencies of RFLP minor alleles in Leith and major alleles only in Cramond showed no relationship with any of the other single variables.

The interrelationships of CHD risk factors and RFLP are summarised in table 14.

Multivariate Analysis

In this analysis (13) evidence of CHD was regressed against a number of individual risk factors simultaneously. The factors included in the analysis were age, hypertension, smoking, (not shown in the table) socio-economic group, positive family history, total / HDL cholesterol ratio > 4.5, (and apolipoprotein B measurements not shown in the table) and RFLP allele status. The Leith and Cramond data were analysed separately; the RFLP minor allele status (haplotypes b, c and d) was entered in Leith and major alleles only (homozygous for haplotype a) in Cramond. In Leith family history emerged as the strongest factor, the only other significant variable being age. In Cramond hypertension emerged as the only significant factor (tables 15a & b).

Table 14. CHD risk factor interrelationship in Leith and Cramond populations combined

Age	Age								
BP ↑	$+^a$	BP ↑							
Obesity	$+^a$	$+^a$	Obesity						
Smoking	$+^a$	$+^a$	$+^a$	Smoking					
Alcohol	0	$+^a$	$+^a$	$+^a$	Alcohol				
FH	$+^*$	0	0	0	0	FH			
S-EG	0	$+^a$	$+^a$	$+^a$	$+^a$	$+^{a*}$	S-EG		
CHD	0	0	0	0	0	$+^{a*}$	$+^a$	CHD	
Lipids	$+^c$	$+^b$	0	0	0	$+^{b*}$	$+^a$	0	Lipids
Genotype †	0	0	0	0	0	0	0	0	0

Lipids = Total chol, HDL chol, Total chol / HDL chol, ApoB, ApoB / ApoAI.
Non fasting Triglyceride levels were related to age[a], obesity[a], alcohol[a].
Hypertriglyceridaemia (fasting TG levels > 2.2mmol/l) were related to obesity[a], smoking[a], alcohol[a], social group[a], hypertension[a] and family hist of CHD[a].
Number of sibs was related to social group[a], smoking[a], alcohol[a], family hist of CHD[a].

* In Cramond, these do not reach p = 0.05	a Chi squared test
† Minor alleles in Leith, Major alleles only in Cramond	b t-test and Mann-Whitney
+ p value at least < 0.01	c Analysis of Variance.
BP↑ Hypertension	S-EG Socio-Economic Group.

Table 15. Estimated partial regression coefficients obtained from multiple logistic regression analysis of evidence of CHD on family history (FH), age, total cholesterol / HDL cholesterol > 4.5 (chol), hypertension (BP), socio-economic group (S-EG) and apo AI/CIII/AIV RFLP genotype including one or more minor allele (a) in Leith and homozygous for major alleles at all four sites (b) in Cramond

Variable	Regression Coefficient	Standard Error	Z	p-value
(a)				
FH	2.369	0.258	3.337	< 0.0005
age	1.044	0.158	2.689	0.0036
chol	0.999	0.108	-0.841	NS
B.P.	0.641	0.465	-0.958	NS
S-EG	1.074	0.122	0.568	NS
RFLP	1.153	0.253	0.564	NS
(b)				
FH	1.401	0.292	1.155	NS
age	0.989	0.185	-0.607	NS
chol	0.987	0.127	-0.100	NS
B.P.	2.765	0.421	2.416	0.008
S-EG	1.028	0.163	0.167	NS
RFLP	1.406	0.277	1.228	NS

Because of the inter-relationship of CHD and RFLP status, the products of these two variables i.e. positive family history and minor alleles (haplotypes b, c and d) in Leith, and positive family history and major alleles only (homozygous for haplotype a) in Cramond, were entered as single variables in the two analyses. The product of FH and RFLP status emerged as the strongest factors in both communities followed again by age in Leith and by hypertension in Cramond. All other factors including family history on

Table 16. Estimated partial regression coefficients obtained from multiple logistic regression analysis of evidence of CHD on the variables listed in table 14 and the combined variable of RFLP and FH (a) in Leith (b) in Cramond

Variable	Regression Coefficient	Standard error	Z	p-value
(a)				
FH	1.511	0.353	1.170	NS
FH × RFLP	2.553	0.513	1.828	0.034
RFLP	0.689	0.381	-0.980	NS
age	1.042	0.158	2.593	< 0.005
chol	0.998	0.109	-0.225	NS
B.P.	0.621	0.468	-1.020	NS
S-EG	1.057	0.122	0.457	NS
(b)				
FH	0.735	0.437	-0.705	NS
FH × RFLP	3.611	0.597	2.150	0.016
RFLP	0.913	0.341	-0.268	NS
age	0.993	0.187	-0.364	NS
chol	0.966	0.128	-0.269	NS
B.P.	2.763	0.425	2.389	0.008
S-EG	1.021	0.166	0.125	NS

its own did not emerge as significant in either analysis (tables 16a & b).

Discussion

There is considerable epidemiological evidence that a history of CHD before age 60 in close relatives is an important risk factor for CHD in middle aged men. CHD attributable to this factor, referred to in this paper as "familial CHD", includes familial clustering or aggregation of several affected subjects but as defined in epidemiological studies it requires only one first degree relative to be affected. In the two samples of North Edinburgh men included in our studies evidence of familial CHD could be demostrated by univariate analysis in Leith but not in Cramond. The difference might be accounted for by a larger number of Leith men reporting CHD in first degree relatives because of a higher rate of maternal and sibling CHD in Leith. The more frequent occurrence of CHD in siblings can be partially accounted for by a significantly larger sibship size in Leith. When both samples were analysed for allele status at four RFLP within or in close proximity to the tightly linked apo AI, CIII and AIV genes, four separate groups were identified. One was a group consisting of men homozygous for the common (or major) allele at all four sites, and the other three were homozygous or heterozygous for a less common (or minor) allele. In an univariate analysis, familial CHD in Leith was restricted to those men who belonged to one or more (double heterozygote) of the minor allele groups. In Cramond, on the other hand, although there was no evidence of familial CHD in the total sample it could be demonstrated in those who were homozygous for major alleles at all four RFLP. Difference in sibship size was not sufficient to account for the discrepancy in familial CHD frequency between the two samples. In a multivariate analysis in which evidence of CHD was regressed simultaneously on age, hypercholesterolaemia, hypertension, socio-economic grouping, family history and RFLP grouping, family history emerged as significant in the total Leith sample but not in Cramond. When family history in the presence of a minor allele was entered as a single covariant in the equation for Leith, the combination emerged as the strongest significant variant and family history on its own was no longer significant. In Cramond when family history and homozygosity for all four major alleles was entered in the equation it also emerged as a strong significant covariant. The variations in nucleotide sequence which account for the four apo AI/CIII/AIV RFLP are all in non coding regions and they are not likely to affect gene function, but by being in close proximity they could serve as markers of functional sequences. The Cramond results suggest that sequences which predispose to CHD are linked to the major alleles of the four RFLP, since in this population those who are homozygous for this haplotype have a significant component of CHD attributable to a family history. The results in Leith also indicate that a CHD determining locus is linked with the same RFLP but in different linkage phase. According to current concepts, the genetic component of CHD liability is due to several genes and CHD results when their additive effect in a given environment exceeds a threshold value (14). We conclude from our findings that the region of the apo AI/CIII/AIV genes is involved in the genetic determination of CHD liability

Familial CHD could be due, at least in part, to the influence of an environment shared in childhood and adolescence with first degree relatives. In both our samples, more so in Leith than Cramond, there were several recognised CHD risk factors such as smoking, diet induced hyperlipidaemia, and obesity, which could be attributable to a common familial environment. Close relatives are also likely to be in the same socio-economic group which, in an univariate analysis, we found to be related to hypertension, obesity, cigarette smoking, alcohol consumption, hyperlipidaemia, CHD and a family history. However in the multivariate simultaneous logistic regression analysis the combined effect on CHD frequency of a family history and RFLP status was independent of socio-economic circumstances and all other CHD risk factors. Moreover in a multivariate analysis of the Leith data which included only the three covariates, family history, minor allele RFLP grouping, and the two combined, the residual effect of family history on evidence of CHD was not significant. These findings suggest that in both populations the influence of a familial environment did not make a significant contribution to familial CHD.

Evidence of a genetic contribution to CHD may be influenced by genotype environment interaction, i.e. the differential response of a given genotype to a different environment. The two samples in our study differed significantly in respect of almost every CHD risk factor examined, hypertension being the only exception. The Cramond sample was slightly older and the Leith men were at higher risk on account of cigarette smoking, hyperlipidaemia, obesity, alcohol consumption and lower socio-economic status. None of these emerged as a significant covariant in the logistic regression analyses in either sample so it is unlikely that they would account for a difference in frequency of familial CHD between the samples. The difference (15.5% of all CHD in Leith and 10.3% of CHD in Cramond) could be accounted for by the greater frequency of maternal and sibling CHD in Leith, the higher rate of sibling CHD being due to bigger sibship size. No evidence of apo AI/CIII/AIV genotype environment interaction was provided therefore by this study.

The results of the Edinburgh studies have provided evidence that the apo AI/CIII/AIV gene region makes a significant contribution to familial CHD. Because the linkage is with different RFLP haplotypes, not only in the two samples but in different sections of the Leith sample, it is likely that case control studies of allele or haplotype frequencies in subjects drawn from these two populations would have given inconclusive or conflicting results. For the same reason the RFLP can not be used to identify individual susceptibility to CHD.

Although there was a higher CHD mortality rate in Leith in preceding years there was only a small and non significant difference between the frequencies with which evidence of CHD was obtained by questionnaire and electrocardiography in the two samples we studied. This probably reflects the poor sensitivity of these methods for detecting a disease which is latent for decades of development before resulting in CHD morbidity and mortality. The figures are likely to have underestimated the true prevalence of the disease and larger numbers might have shown a significant difference between the two populations. Diagnosis by questionnaire and electrocardiography has however proved highly specific in large scale studies in which subjects with evidence of CHD obtained by these methods have been shown to have a greatly increased probability of subsequent CHD events (15). In other RFLP studies different criteria of CHD diagnosis have been used e.g. acute myocardial infarction and angiographic evidence of coronary atherosclerosis. Myocardial infarction is a highly selected category of CHD and coronary atherosclerosis is not synonymous with CHD. Fatal CHD events may occur in patients with minimal angiographic abnormality and subjects with extensive coronary artery disease may never suffer from CHD. The use of different methods to diagnose CHD is another possible reason for inconsistency in the published reports of CHD association with apo AI/CIII/AIV RFLP.

We conclude from these RFLP studies that the apo AI/CIII/AIV gene region is implicated in the determination of premature CHD.

The work described in this paper was carried out with SW Morris, PR Wenham, Mrs JL Lowther, Mrs. PRS McKenzie and in collaboration with AG Donald and partners. Statistical advice was provided by Dr. RA Elton. It was supported by the McKenzie trust and the Margaret Kennedy endowment fund.

References

1.   Neufeld HN, Goldbourt U. Coronary Heart Disease: Genetic Aspects. *Circulation* 1983; **67:** 943-54.

2.   Goldbourt U, Neufeld HN. Genetic Aspects of Arteriosclerosis. *Arteriosclerosis* 1986; **6:** 357-77.

3.   Hopkins PN, Williams RR. Human Genetics and Coronary Heart Disease: A Public Health Perspective. *Annu Rev Nutr* 1989; **9:** 303-45.

4.   Karathanasis SK, Norum RA, Zannis VI, Breslow JL. An Inherited Polymorphism in the Human Apolipoprotein AI Gene Locus related to the Development of Atherosclerosis. *Nature* 1983; **301:** 718-20.

5.   Rees A, Shoulders CC, Stokes J, Galton DJ, Baralle FE. DNA Polymorphism adjacent to Human Apo-Protein A-I Gene. Relation to Hypertriglyceridaemia. *Lancet* 1983; *i:* 444-6.

6.   Ferns GAA, Galton DJ. Haplotypes of Human Apolipoprotein AI-CIII-AIV Gene Cluster in Coronary Atherosclerosis. *Hum Genet* 1986; **73:** 245-9.

7. Breslow JL. Apolipoprotein Genetic Variation and Human Disease. *Physiol Rev* 1988; **68:** 85-132.

8. Fisher EA, Coates PM and Cortner JA. Gene Polymorphisms and Variability of Human Apolipoproteins. *Annu Rev Nutr* 1989; **9:** 139-60.

9. Sing FC, Moll PP. Genetics of variability of CHD risk. *Int J Epidemiol* 1989; **18:** 5183-95

10. Price WH, Morris SW, Kitchin AH, Wenham PR, Burgon PRS and Donald PM. DNA restriction fragment length polymorphisms as markers of familial coronary heart disease. *Lancet* 1989; *i:* 1407-1410

11. Price WH, Morris SW, Kitchin AH, Wenham PR, Lowther JL, McKenzie PRS and Donald PM. Further evidence implicating the apo AI/CIII/AIV gene region in genetic liability to CHD. In preparation.

12. Kunkel LM, Smith KD, Boyer SH, et al. Analysis of Y Chromosome Specific Reiterated DNA in Chromosome Variants. *Proc Natl Acad Sci USA* 1977; **74:** 1245-49.

13. Lee J. Covariance adjustment of rates based on the multiple logistic regression model. *J Chron Dis* 1981; **34:** 415-426

14. Falconer DS. The inheritance of liability to certain diseases, estimated from the incidence among relatives. *Ann Hum Gen* 1965; **29** 51-62

15. Shaper AG, Pocock SJ, Walker M, Phillips AN, Whitehead TP, Macfarlane PW. Risk factors for ischaemic heart disease: the prospective phase of the British regional heart study. *J Epid Comm Health* 1985; **39:** 197-209

DNA POLYMORPHISMS AS DISEASE MARKERS

David J. Galton and John C. Alcolado

Department of Human Metabolism and Genetics
Medical Professorial Unit
St. Bartholomew's Hospital
London EC1A 7BE   U.K.

INTRODUCTION

One of the surprizing observations arising from recent sequencing studies of the human genome is the large amount of nucleotide variation that is found in intergenic and intronic DNA. As many as 1:300 bases may be variable (1). This creates a wealth of new genetic markers that can be used to track almost any locus in pedigree or population studies. Clinical genetics has been transformed from a study of pedigrees in which rare mutant proteins co-segregate with affected members to a study of any inherited disease whose genetic variance exceeds approx. 0.65 (2). DNA variation can be used to attempt to locate a possible aetiological locus in such diseases as premature atherosclerosis, non-insulin dependent diabetes mellitus (Type 11), the hyperlipidaemias, essential hypertension and the pyscho-affective disorders such as schizophrenia or the manic-depressive psychoses.   Instead of mapping the aetiological locus by studying recombinational events in pedigrees, it can now be located by the use of overlapping ordered DNA clones in genomic libraries. The function of the gene can be deduced from its DNA sequence and the type of protein it encodes rather than doing tedious complementation studies of affected and unaffected cells such as skin fibroblasts in tissue culture.

DNA Polymorphisms

Three main types of DNA polymorphisms are in current use: restriction fragment length polymorphisms (RFLPs); variable number of tandem repeats (VNTRs); and (CA)n minirepeats (3).

RFLPs were the first to be used and are produced by nucleotide alterations that either create or abolish sites for the action of restriction endonucleases. Only alterations that affect the activity of restriction endonucleases can be detected by this method, and it is likely that much more nucleotide variation occurs than can be revealed by RFLP's. The polymorphism usually produces two alleles which may not be very helpful in distinguishing parental chromosomes, particularly if one of the alleles occurs rarely in the population. Several RFLP's at the same locus can be used to construct haplotypes and these are often much more

DNA Polymorphisms as Disease Markers, Edited by D.J. Galton and
G. Asmann, Plenum Press, New York, 1991

149

informative than single site polymorphisms. However the work involved in their identification and the difficulties involved in the analysis of sets of haplotypes (by constructing cladograms) often precludes their use.

VNTRs are due to short sequences of about 4 - 6 nucleotides (often GC-rich) repeated in variably sized blocks of DNA and can provide very useful genetic markers. In some examples such as the VNTR at the 5'-prime end of the insulin gene on chromosome 11, more than 70% of individuals are heterozygous for length (and probably more for DNA sequence) at this hypervariable polymorphic site (4). Unambiguous identification of the transmission of parental chromosomes can be achieved in many cases. One possible disadvantage of VNTR's is their instability in that they were probably produced by a series of unequal recombinational events during meiosis. If the sites are very unstable with regard to cross-overs, their value as genetic markers will be correspondingly reduced, particularly if one is studying an aetiological mutation that occurred many years ago in our ancestral gene pool.

A third class of DNA polymorphisms are the minirepeats of nucleotides (CA)n widely scattered throughout the entire genome, approximately 30 kb apart, with little tendency to form clusters (5). At least twelve of these repeats have been tested for length polymorphism and were found to have an average polymorphism information content (PIC) value of 0.55, and were inherited in a normal co-dominant fashion. (CA) repeats can be identified by screening cosmid or bacterophage human genomic DNA libraries with poly (dG-dT), poly (dA-dC) probes; and those close to candidate genes provide very useful genetic markers. When their flanking sequences are determined, they can be amplified by the PCR reaction to determine the variable size of the allelic fragments and hence used as linkage markers(6). Such types of polymorphic loci and potentially provide linkage markers for any candidate gene that may be implicated in disease production. Their value in monogenic diseases such as cystic fibrosis, Huntington's chorea, Duchenne Muscular Dystrophy have been well-established. The general strategy of assembling two or three generation pedigrees with many affected members; performing co-segregation studies with polymorphic linkage markers; assessing strengths of association by LOD scores; and mapping the aetiological locus with even closer markers has been well-validated. The strategy for multifunctional or polygenic disease is less well defined. The assumption is made that a few major loci are contributing to the genetic variance of the disease with perhaps several minor genes contributing to a background effect. Pedigree studies with LOD-score analyses have never been shown to be of use to locate an aetiological locus for a polygenic disorder. Disease association studies in populations have however been successful in identifying an apolipoprotein E polymorphism as a locus contributing to the genetics of Type III hyperlipidaemia. The HLA B8 locus, in linkage disequilibrium with the DQ allele of the DR locus, was found by population studies to be implicated in the genetics of insulin dependent diabetes mellitus. The conceptual overview for the stragegy to elucidate polygenic disease is shown in Fig 1.
The trait can either be a disease phenotype or a population parameter such as the level of blood cholesterol or triglyceride. The disease phenotype can either be an end-stage pathology such as premature atherosclerosis, or perhaps more appropriately an intermediate phenotype such as a familial hyperlipidaemia (that in

CONCEPTUAL OVERVIEW

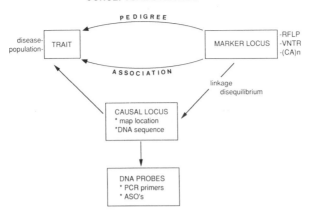

Figure 1. Conceptual Overview

turn predisposes to premature atherosclerosis). The polymorphic
marker loci can be as previously described (either as RFLPs,
VNTRs; or a (CA)n mini-repeats) and their association with the
trait can be tested by either population or pedigree genetics. In
our laboratory we use the population approach to screen loci to
determine if they could be implicated in the genetics of the
disease. Although there are numerous pitfalls (particularly in the
ethnic matching of the control group, and the possible
heterogeneity of the patient group), disease association studies
have been of proven value in the preliminary identification of a
genetic loci for Type III hyperlipidaemia and insulin dependent
diabetes mellitus. Thereafter a most useful pedigree approach is
to use affected sib-pair analysis (Fig 2   ).

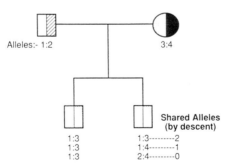

Figure 2. Affected Sib-Pair Analysis

In this case only affected sib-pairs of a pedigree are considered, the other relatives are discarded from the analysis. This is particularly useful in diseases of late-onset when it is not clear if an unaffected relative may go on to develop the disease; or if the diagnosis is in doubt in an individual , he can be rejected from the final calculations. Other advantages of this approach are that no assumptions need to be made on the mode of transmission of the disease (i.e. whether it is a Mendelian, dominant or recessive etc), which is essential for linkage analysis; also no control group is required since the results are based on the extent of concordance of shared haplotypes (Fig 2). In the example of Fig 2, 25% of sib pairs would be expected to share identical alleles by chance (from the laws of meiosis and random segregation). Fifty percent of sib-pairs would be expected to share one haplotype in common; and 25% would be non-identical. If there is sib-pair distortion of these ratios, it suggests that this locus may be contributing to the genetics of the disease phenotype. No absolute association of haplotype with disease would be expected because of variable environmental factors, heterogeniety of the disease and recombination events disturbing the haplotype-disease relationship.

Another ingenious pedigree approach is to study mono - versus di-zygotic twins for the disease trait to see if particular haplotypes at a locus influence intra-pair twin differences. In some cases a particular haplotype may be associated with very restricted mono-zygotic twin-pair differences, whereas other haplotypes may show quite marked variability (of twin-pair differences). Again this would imply that the locus is contributing to the genetic variance of the phenotype.

Finally having identified a reliable marker locus, the next stage in the analysis is to attempt to identify the causal locus by selecting ever-closer genetic markers. However the reliability of the initial marker must be thoroughly tested on several groups of patients preferably from widely differing populations before proceeding to an aetiological analysis. There are examples where published evidence for a genetic marker has had to be retracted mainly due to problems of inter-breeding immigrant populations of North America where the relationships of linkage disequilibrium that associate marker to causal locus can be distorted (Fig 1). It is very important to select both homogeneous patient and control groups for such studies.

Disease Markers

The best established example for polygenic disease is the involvement of the HLA locus on chromosome six and insulin dependent diabetes. This was initially observed from disease association studies with HLA B8 which led to the recognition of involvement of the HLA.DR3/DR4 locus (7). The latter haplotype is present in 95% of diabetics compared to approximately 40% in controls. Further analysis has suggested that mutation of a residue 57 (aspartate) of the DQ gene may play a crucial role in the aetiology of insulin dependent diabetes mellitus (8).

In the field of lipids and atherosclerosis the best studied example is the apolipoprotein E polymorphism and Type III hyperlipidaemia, particularly as the polymorphic marker sites were found to be the aetiological sites. However the prevalence of Type III hyperlipidaemia is only approximately 1:5000. A more common disorder is hypertriglyceridaemia / low HDL and Table 1 summarises the disease association studies that have been performed with polymorphic sites at the apo AI/CIII/AIV gene cluster on chromosome 11.

TABLE 1

CASE-CONTROL STUDIES OF DNA POLYMORPHISMS
AT THE APO AI/CIII/AIV GENE CLUSTER AND
HYPERLIPIDAEMIA / ATHEROSCLEROSIS

Senior Authors	Journal	Sites	Significant Association	
			Lipids	Atheroscl.
1. Galton, Baralle	Lancet '83	Sst-1	<0.01	...
2. Baralle, Galton	J.C.Invest '85	Sst-1	<0.01	...
3. Rees, Galton	Atheroscleros.'85	Sst-1	...	<0.02
4. Ferns, Galton	Lancet '85	Sst-1	...	<0.01
5. Kessling, Humphries	Clin Genetics '85	Xmn-1	<0.02	...
6. Deeb,Hayden	Unpublished data	Sst-1	<0.01	...
7. Henderson, Berger	Hum. Gen '87	Sst-1	<0.02	...
8. A-Setala, Nikkila	Atheroscleros.'87	Sst-1	<0.001	no
9. Hayden, Hewitt	Am.J.Genetics '87	Xmn-1 Sst-1	<0.03	...
10. Paulweber, Sandhofer	Atheroscleros.'88	Xmn-1 Pst-1 Pvu-II	<0.01	no
11. Satoh, Hamaguchi	Jpn.J.Hum.Genet.'87	Sst-1 Msp-1	...	<0.05
12. Price, Morris	Lancet 1989	Xmn-1 Pst-1 Msp-1 Sst-1	...	<0.005
13. Cole, Ferrell	Am.J.Hum.Gen '89	Xmn-1 Sst-1	<0.05	...
14. Shoulders, Baralle	Atheroscler. '89	Sst-1	<0.001	...
15. Myant, Thompson	Atheroscler. '89	Pst-1 Xmn-1	<0.05	<0.02

The populations studied range from Northern and Central Europe, Scandinavia, North America and Japan. The lipid abnormalities (of raised plasma triglyceride and/or low HDL) may be considered as an intermediate phenotype to premature atherosclerosis; and the data are more consistent with the former trait.

Attempts have been made to go from the polymorphic markers at the apo AI/CIII/AIV gene cluster to an aetiological locus. One possibility is a common nucleotide substitution at position - 78 in the promoter region of the apolipoprotein AI gene on chromosome 11. This mutation alters the ability of the promoter to drive expression of the gene for chloramphenicol acetyl transferase when transfected into Hep G2 cells or HELA cells (in the CAT assay).

The frequency of the A-allele as detected by hybridisation with allele specific oligonucleotides is also different in the top and lowest decile of the distribution curre for HDL in groups of healthy females (9). So this promoter mutation may be affecting the expression of the apo Al gene _in vivo._ Further work along these lines, as well as exploration of other critical loci, such as the lipoprotein lipase gene on chromosome 8p 22 (10) is required to clarify the genetic determinants of the hypertriglyceridaemia / low HDL syndrome.

Conclusions

One can confidently predict that the subject of DNA polymorphisms as disease markers will develop along the lines of the HLA field and provide a wealth of new genetic markers for the inherited basis of all the common metabolic diseases. In some instances the markers will lead to identification of aetiological loci and provide valuable insights into the pathogenesis of the condition. Alternative modes of therapy may be suggested from the gene products of such incriminated loci. Finally some markers will become of diagnostic value, as is HLA B27 locus for the diagnosis of ankylosing spondylitis, even though the causal relationships are not fully understood. Since various types of DNA polymorphisms are scattered regularly throughout the entire genome, the potential for investigation of any suspected locus as a genetic determinant for disease remains an exciting prospect.

Acknowledgements

The authors gratefully acknowledge financial support from the MRC (UK) for David J Galton and a Lawrence Research Fellowship of the British Diabetic Association to John C Alcolado.

References

1. Jeffreys AJ, 1979. DNA sequence variants in the C , A , delta and beta globin genes of man. Cell 18.1
2. Galton DJ, 1985. "Molecular Genetics of Common Metabolic Disease." Publ. E Arnold, London.
3. Litt M, Luty JA, 1989. A Hypervariable micro-satellite revealed by _in vitro_ amplification of dinucleotide repeats within the actin gene. Am. J. Hum Genet. 44, 397.
4. Hitman GA, Tarn AC, Winter RM, Bottazzo F, Galton DJ, 1985. Type I diabetes and a highly variable locus close to the insulin gene on chromosome 11. Diabetiologia 28, 218.
5. Weber JL, May PE, 1989. Abundant class of human DNA polymorphisms which can be typed using PCR. Am. J. Human Genet. 44, 388.
6. Carle GF, Frank M, Olson MV, 1986. Electrophoretic separation of large DNA molecules by periodic inversion of the electric field. Science 232, 65.
7. Cudworth AG, Woodrow JC, 1975. Evidence of HLA-linked genes in juvenile diabetes m. Br Med J II, 133.

8. Todd JA, Bell JI, McDevitt HO, 1987. HLA-DQ gene contributes to susceptibility and resistance to IDDM. Nature 329, 599.

9. Sidoli A, Galliani S, Baralle FE, 1990. DNA based diagnostic tests: recombinant DNA and cardiovascular disease risk factors in "Lipids and Cardiovascular Disease." ed. Galton DJ, Thompson GR. Publ. Churchill Livingstone, Edinburgh, UK.

10. Chamberlain JC, Thorn JA, Oka K, Galton DJ, Stocks J, 1989. DNA polymorphisms at the lipoprotein lipase gene: associations in normal and hypertriglyceridaemic subjects. Atheroscler. 79, 85.

11. Rees A, Stocks J, Shoulders CC, Galton DJ, Baralle FE, 1983. DNA polymorphisms adjacent to the human apolipoprotein AI gene in relation to hypertriglyceridaemia. Lancet I, 444.

12. Rees A, Stocks J, Sharpe CR, Shoulders CC, Baralle FE, Galton DJ, 1985. DNA polymorphism in the apo AI/CIII gene cluster: association with hypertriglyceridaemia. J. Clin Invest. 76, 1090.

13. Rees A, Williams LG, Caplin J, Stocks J, Camm J, Galton DJ 1985. DNA polymorphisms flanking the insulin and apo C-III genes and atherosclerosis. Atheroscler. 58, 269.

14. Ferns GA, Stocks J, Galton DH, 1985. Genetic polymorphisms of apolipoprotein CIII and insulin in survivors of myocardial infarction. Lancet I, 300.

15. Kessling AM, Horsthemke B, Humphries SE, 1985. A study of DNA polymorphisms around the human apolipoprotein AI gene in hyperlipidaemic and normal individuals. Clinical Genetics 28, 296.

16. Deeb S, 1990 in 'DNA Polymorphisms as Disease Markers'. ed. Galton DJ, Assmann G.(in press).

17. Henderson HE, London SV, Michie J, Berger GMB, 1987. Association of a DNA polymorphism in the apo CIII gene with diverse hyperlipidaemic phenotypes. Human Genet. 75, 62.

18. Aalto-Setala K, Kontula K, Sane T, Nikkila E 1987. DNA polymorphisms of the apo AI/CIII and insulin genes in familial hypertriglyeridaemia and coronary heart disease. Atheroscler. 66, 145.

19. Hayden MR, Kirk H, Clark C, Frohlich J, McLeod R, Hewitt J 1987. DNA polymorphisms around the apo AI-CIII genes and genetic hyperlipidaemias. Am J Hum. Genet. 40, 421. Paulweber B, Friedl W, Krempler F, Humphries SE, Sandhofer F 1988. Genetic Variation in the apo AI-CIII-AIV gene cluster and coronary heart disease. Atheroscler. 73.

21. Satoh J, Hattori N, Oniki M, Yamakawa K et al 1987. Apo-AI-CIII gene polymorphisms in Japanese myocardial infarction survivors. Jpn J. Hum. Genet. 32, 15.

22. Price WH, Morris SW, Kitchin AH, Wenham PR, Burgon PRS, Donald PM, 1989. DNA restriction fragment length polymorphisms as markers of familial coronary heart disease. Lancet I, 1407.

23. Cole SA, Szathmary EJE, Ferrell RE, 1989. Gene variation in the apo AI/CIII/AIV cluster in Dogrib Indians of the Northwest Territories. Am. J. Hum. Genet. 44, 835.

24. Shoulders CC, Ball MJ, Baralle FE, 1989. Variation in the apo AI/CIII/AIV gene complex: its association with hyper-lipidaemia. Atheroscler. 80, 111.

25. Wile DB, Barbir M, Gallagher J, Myant NB, Thompson GR, Humphries SE, 1989. Apo AI gene polymorphisms: frequency in patients with coronary artery disease, healthy controls and association with serum apo AI and HDL-concentrations. Atheroscler. 78, 9.

PARTICIPANTS

J. ALCOLADO   Department of Medicine, St. Bartholomew's Hospital
West Smithfield, London EC1A 7BE, UK

D. AMEIS   Department of Medicine, Hamburg, FRG

G. ASSMANN   Director, Zentrallaboratorien der Medizinische
Einrichtungen, University of Munster, FRG

F.E. BARALLE   Institute Sieroterapico, Milan, Italy

K. BERG   Institute of Medical Genetics, Oslo, Norway

E. BOERWINKLE   Graduate School, Biochemical Sciences, Houston, Texas, USA

J.C. CHAMBERLAIN   Department of Pathology, Royal Free Hospital, Pond Street
Hampstead, London NW3, UK

L. CHAN   Baylor College of Medicine, Houston, Texas, USA

K. DAVIES   Institute of Molecular Medicine, Oxford, UK

S. DEEB   Department of Medicine (Genetics) Seattle, Washington, USA

H. FUNKE   Institute for Clinical Chemistry, Munster, FRG

C. GABELLI   Department of Medicine, Padova, Italy

D.J. GALTON   Department of Medicine, St. Bartholomew's Hospital
West Smithfield, London EC1A 7BE, UK

L.M. HAVEKES   TNO Gabius Institute, Leiden, The Netherlands

S. HUMPHRIES   Charing Cross Sunley Research Centre, London, UK

A. MOTULSKY   Washington University, Department of Medical Genetics
Seattle, Washington, USA

M. MUECKLER   Department of Cell Biology, Washington School of Medicine
St. Louis, Missouri, USA

J.M. ORDOVAS   Tufts University, Boston and The Institute of Biochemistry
Zaragoza, Spain

A. PERMUTT   Washington University School of Medicine
St. Louis, Missouri, USA

W.H. PRICE   Department of Medicine, Western General Hospital
                  Edinburgh, EH4 2XU, UK

J. SCOTT   Clinical Research Centre, Harrow, Middlesex, UK

C. SIRTORI   Institute of Pharmacology, Milan, Italy

J.A. THORN   Department of Medicine, St. Bartholomew's Hospital
                  West Smithfield, London EC1A 7BE, UK

G. UTERMANN   Institute of Medical Biology and Genetics
                  Innsbruck, Austria

INDEX

Haplotypes, 21
Hemochromatosis, 4
Hepatic triglyceride lipase, 62
Heritability, 2
Heterozygotes, 3
High density lipoprotein (HDL), 61, 91
HLA B27, 3
Hyperapobetalipoproteinaemia, 79, 87
Hypertriglyceridaemia, 111 - 121
Hypertriglyceridaemia/low HDL, 152
Hypervariable region, 54

Insulin dependent diabetes, 152,
 Insulin gene, 44 - 47
  polymorphisms, 45
  promoter, 53
Insulin receptor gene, 47 - 50
Intermediate phenotypes, 2 - 3

Kringle 4, 127

LCAT activity, 64
  gene, 66
  deficiency, 66 - 67
LDL receptor gene, 5, 129
LDL synthesis, 84
Leith, 142,
Linkage analysis, 3, 6, 45,
  markers, 123 - 124, 150
Lipid transport, 7
Lipoprotein lipase (LPL), 105 - 106, 112, 130
  gene variants, 111 - 121
  genotypes, 116 - 121
Lp (a) lipoprotein, 126 - 127
Lipoprotein-modifying genes, 7

Major genes, 6
Mendelian genes, 1
Monozygotic twins, 2
Multifactorial diseases, 2
Multiple genes, 1

Na/glucose co-transporters, 28
NIDDM, 36, 43,
NOD mouse, 4

Pima Indians, 53
Plasma apolipoproteins, 6
Polygenic disease, 15
Polymerase chain reaction (PCR), 53
Public health implications, 12

Restriction fragment length
polymorphisms (RFLPs), 3, 135, 149
  markers, 135 - 147
Risk factors, 15
Risk prediction, 2

Segregation analysis, 3
Single strand confirmation
polymorphism, 53
Susceptibility genes, 123

Twin studies, 2
Type III hyperlipoproteinemia, 71
152

Variable number of tandem repeats,
(VNTRs), 3, 150
VLDL, 71